BUILDING

CODE REFERENCE

Third Edition

Written by
American Contractors Exam Services

Published by
CENGAGE Learning

www.DeWALT.com/GUIDES

DEWALT® Building Code Reference: Based on the 2015 International Residential Code, Third Edition
American Contractors Exam Services

SVP, GM Skills & Global Product Management: Dawn Gerrain

Product Director: Matthew Seeley

Senior Product Manager: Vanessa Myers

Senior Director, Development: Marah Bellegarde

Senior Product Development Manager: Larry Main

Associate Content Developer: Jenn Alverson

Product Assistant: Jason Koumourdas

Vice President, Marketing Services: Jennifer Ann Baker

Marketing Manager: Scott Chrysler

Senior Production Director: Wendy Troeger

Production Director: Andrew Crouth

Senior Content Project Manager: Stacey Lamodi

Managing Art Director: Jack Pendleton

Library of Congress Control Number: 2015960338

ISBN: 978-1-305-66699-3

Cengage Learning
20 Channel Center Street
Boston, MA 02210
USA

Cengage Learning is a leading provider of customized learning solutions with employees residing in nearly 40 different countries and sales in more than 125 countries around the world. Find your local representative at **www.cengage.com.**

Cengage Learning products are represented in Canada by Nelson Education, Ltd.

Printed in the United States of America
Print Number: 01 Print Year: 2016

To learn more about Cengage Learning, visit **www.cengage.com**

Purchase any of our products at your local college store or at our preferred online store
www.cengagebrain.com

Notice to the Reader

Publisher and DEWALT® do not warrant or guarantee any of the products described herein or perform any independent analysis in connection with any of the product information contained herein. Publisher and DEWALT® do not assume, and expressly disclaims, any obligation to obtain and include information other than that provided to it by the manufacturer. The reader is expressly warned to consider and adopt all safety precautions that might be indicated by the activities described herein and to avoid all potential hazards. By following the instructions contained herein, the reader willingly assumes all risks in connection with such instructions. The publisher and DEWALT® make no representations or warranties of any kind, including but not limited to, the warranties of fitness for particular purpose or merchantability, nor are any such representations implied with respect to the material set forth herein, and the publisher and DEWALT® take no responsibility with respect to such material. The publisher and DEWALT® shall not be liable for any special, consequential, or exemplary damages resulting, in whole or part, from the readers' use of, or reliance upon, this material.

About the International Code Council®

The International Code Council is a member-focused association. It is dedicated to developing model codes and standards used in the design, build and compliance process to construct safe, sustainable, affordable and resilient structures. Most U.S. communities and many global markets choose the International Codes. ICC Evaluation Service (ICC-ES) is the industry leader in performing technical evaluations for code compliance fostering safe and sustainable design and construction.

Washington DC Governmental Affairs Office:

500 New Jersey Avenue, NW, 6th Floor, Washington, DC 20001-2070

Regional Offices:

Eastern Regional Office (BIR); Central Regional Office (CH);
Western Regional Office (LA)
888-ICC-SAFE (888-422-7233)
www.iccsafe.org

Titles Available From

DEWALT®

Trade Reference

Blueprint Reading

Construction—Master Edition

Construction Estimating

Construction Safety/OSHA

Electric Motor

Electrical Estimating

Electrical

HVAC/R—Master Edition

Plumbing

Plumbing Estimating

Spanish/English Construction Dictionary – Illustrated Edition

Wiring Diagrams

Exam and Certification

Building Contractor's Licensing

Electrical Licensing

HVAC Technician Certification

Plumbing Licensing

Code Reference

Building

Electrical

HVAC

Plumbing

Business Reference

Contractor's Forms & Letters

Contractor's Daily Logbook and Jobsite Reference

Construction Estimating Complete Handbook

Quick Check

Wiring Quick Check

Construction Math Quick Check

CONTENTS

CODE OVERVIEW	
Code	**Description**
R101.3	The purpose of the code is to provide MINIMUM requirements to protect the public and safeguard health.
R102.2	The provisions of the code will not nullify any provisions of local, state, or federal laws.
R104.1	The building official (usually hired by the local municipality) is authorized to enforce the provisions of the code.
R104.2	The building official (typically via the local building department office) will receive applications, review construction documents and issue permits for the erection and alteration of building and structures. He will "enforce" the provisions of the code by inspecting the projects that have been issued permits.
R104.6	The building official has the right to enter premises where code violations are suspected to exist at reasonable times upon request.
R105.1	Permits are required by any owner or authorized agent who intends to construct, enlarge, alter, repair, move, demolish or change the occupancy of a building or structure, or to erect, install, enlarge, alter, repair, remove, convert or replace any electrical, gas, mechanical or plumbing system, the installation of which is regulated by code.

Exempt from Permit (Check with local building official to confirm):

- Detached (1-story) structures used for storage, recreation, etc. if the floor area does not exceed 200 square feet.
- Fences 7' or lower.
- Retaining walls that do not measure more than 4' from the bottom of the footing to the top of the wall (unless supporting a surcharge)
- Water tanks supported directly on grade if the capacity does not exceed 5000 gallons and ratio of height to diameter or width does not exceed 2 to 1.
- Sidewalks and driveways.
- Finish work such as painting, papering, tiling, carpeting, cabinet installation, etc.
- Prefab swimming pools that are less than 24" deep.

YOU SHOULD KNOW: LOCAL CODES

- Check your neighborhood "Restrictive Covenants" to make sure your project is in compliance with the agreement between you and your neighbors.
- Check local laws to make sure permitting is not required for any of the exemptions listed.

PERMITS AND INSPECTIONS	
Code	**Description**
R105.3	To obtain a permit, complete a written application on a form provided by your local building official's office.
R105.3.2	If you do not respond to solicitation for additional information and the permit has not been issued within 180 days, it will be deemed abandoned.
	Work must commence within 180 days of the permit issuance date.
R105.6	The building official can suspend or revoke a permit if it was issued in error or on the basis of incorrect, inaccurate or incomplete information or if it is found to be in violation of code.
R108.1	The local building department will charge a permit fee based on an established schedule of permit fees. The permit will not be valid until the fees have been paid.
R109.1	Permit holder will notify the building official as required for necessary inspections. The building official will either approve that part of the construction as completed or will notify the permit holder if it fails inspection.
R109.1.1	Foundation inspection is to be made after poles or piers are set or trenches/basement areas have been properly excavated and any required forms erected with all necessary reinforcement in place. This inspection will include excavations for the support of bearing walls, partitions, structural supports, or equipment and special requirements for wood foundations.
R109.1.2	Rough inspection of plumbing, mechanical, gas, and electrical systems must be made prior to covering or concealment, before fixtures or appliances are set or installed, and prior to framing inspection.
R109.1.4	Inspection of framing and masonry construction is to be made after the roof, masonry, all framing, firestopping, draftstopping and bracing are in place and after the plumbing, mechanical and electrical rough inspections are approved.
R109.1.5	The building official may make or require additional inspections to ascertain code compliance.
R109.1.6	Final inspection will be made after the permitted work is complete and prior to occupancy.
R110.1	No building or structure can be used or occupied, and no change in the existing occupancy classification of a building or structure or portion thereof shall be made until the building official has issued a certificate of occupancy.

CONSTRUCTION DOCUMENTS

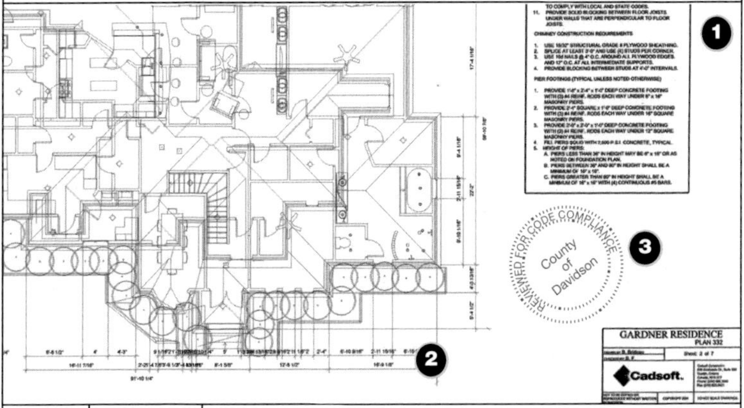

No.	Code	Description
1	R106.1	Construction Documents must be submitted in two or more sets (one to be retained by the Building Official) with each application for permit. If required by the state, the documents must be prepared by a registered design professional. The building official can require additional construction documents to be prepared.
2	R106.1.1	The documents must be drawn on suitable material. Electronic media documents are permitted IF approved by the building official. The documents must be of sufficient clarity to indicate the location, nature and extent of the work proposed and show in detail that it will conform to the provisions of the code, relevant laws, ordinances, rules and regulations as determined by the building official.
3	R106.3.1	Upon issuance of a permit, the documents shall be approved, in writing or by a stamp which states "REVIEWED FOR CODE COMPLIANCE." One set of construction documents will be retained by the building official. The other set must be kept at the site of work and available by inspection to the building official or authorized building official's agent.

YOU SHOULD KNOW: CONSTRUCTION DOCUMENTS

- **R106.5:** One set of approved construction documents will be retained by the building official for a period of at least 180 days from date of completion of the permitted work or as required by state or local laws.
- **R802.10.1:** Truss design drawings must be provided to the Building Official and approved by the building official prior to installation.

SITE (PLOT) PLANS

Building setbacks from lot lines are also regulated by jurisdiction zoning codes.

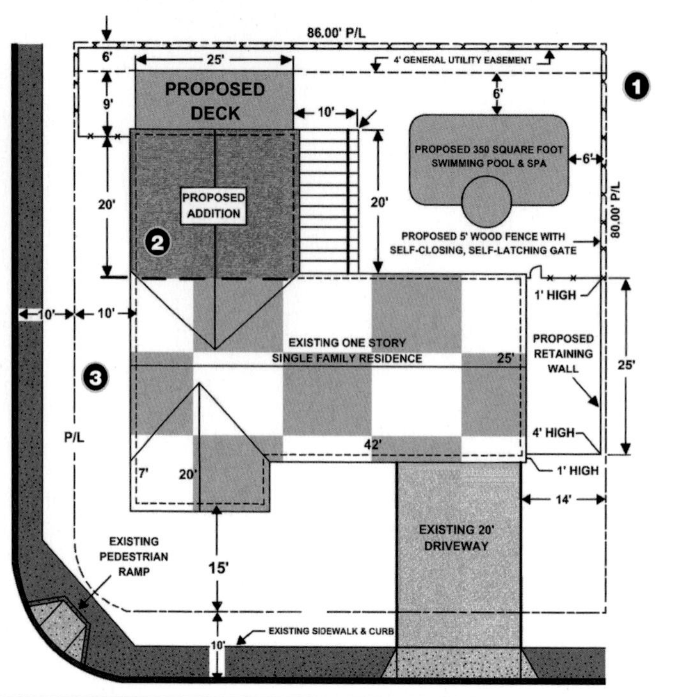

No.	Code	Description
1	R106.2	Construction documents submitted for permits must be accompanied by a site plan.
2	R106.2	Site plan must show the size and location of all new construction and existing structures on the site.
3	R106.2	Site plan must show all distances from lot lines.

YOU SHOULD KNOW:

- Drawings should be drawn to scale with the scale indicated on the drawings.
- Typically, the minimum size of a site plan accepted is $8\frac{1}{2}" \times 11"$.
- Most permitting departments require that all streets be shown and all distances to the property lines must be shown for each corner and side of the structure including any offsets.

FOUNDATION DRAINAGE

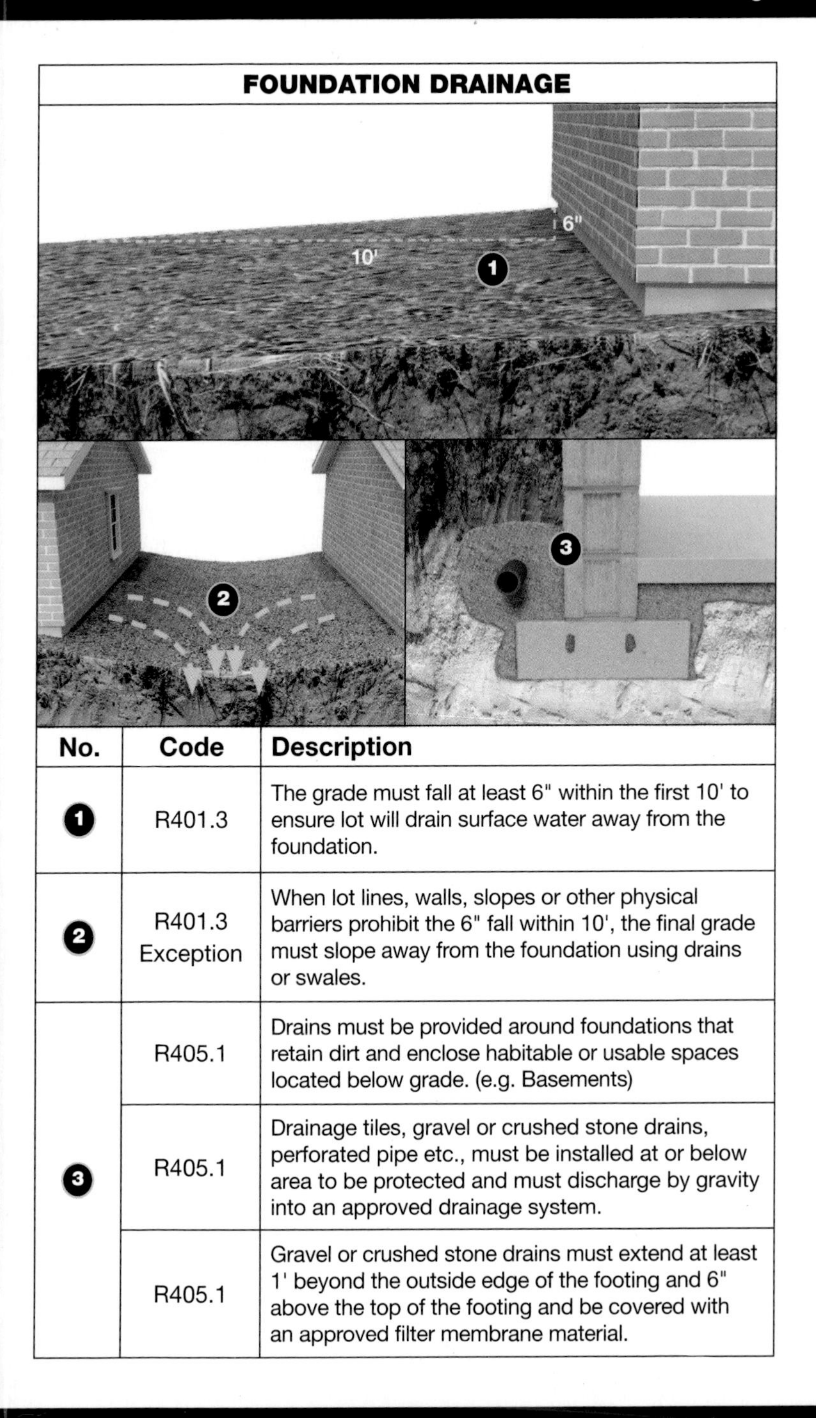

No.	Code	Description
1	R401.3	The grade must fall at least 6" within the first 10' to ensure lot will drain surface water away from the foundation.
2	R401.3 Exception	When lot lines, walls, slopes or other physical barriers prohibit the 6" fall within 10', the final grade must slope away from the foundation using drains or swales.
3	R405.1	Drains must be provided around foundations that retain dirt and enclose habitable or usable spaces located below grade. (e.g. Basements)
	R405.1	Drainage tiles, gravel or crushed stone drains, perforated pipe etc., must be installed at or below area to be protected and must discharge by gravity into an approved drainage system.
	R405.1	Gravel or crushed stone drains must extend at least 1' beyond the outside edge of the footing and 6" above the top of the footing and be covered with an approved filter membrane material.

CEILING HEIGHT

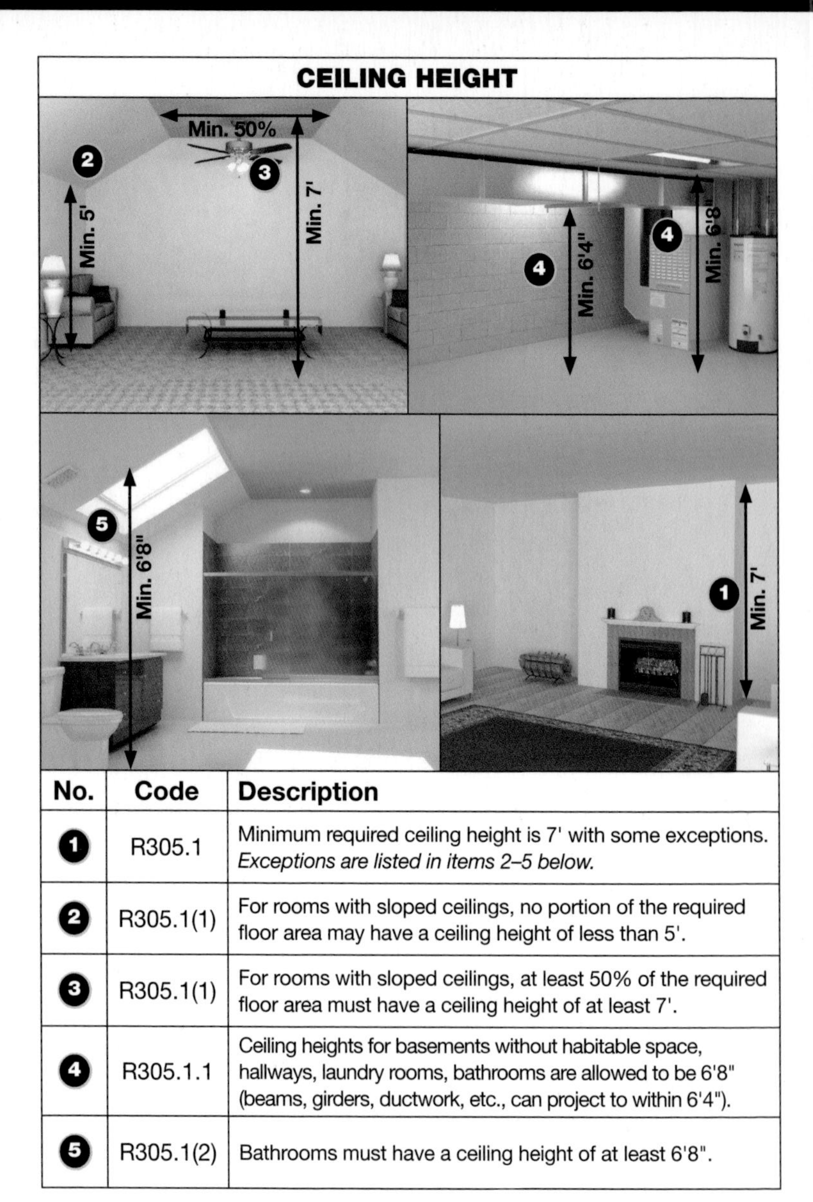

No.	Code	Description
1	R305.1	Minimum required ceiling height is 7' with some exceptions. *Exceptions are listed in items 2–5 below.*
2	R305.1(1)	For rooms with sloped ceilings, no portion of the required floor area may have a ceiling height of less than 5'.
3	R305.1(1)	For rooms with sloped ceilings, at least 50% of the required floor area must have a ceiling height of at least 7'.
4	R305.1.1	Ceiling heights for basements without habitable space, hallways, laundry rooms, bathrooms are allowed to be 6'8" (beams, girders, ductwork, etc., can project to within 6'4").
5	R305.1(2)	Bathrooms must have a ceiling height of at least 6'8".

YOU SHOULD KNOW:

- Required height shall be measured from the finished floor to the finished ceiling.
- Habitable Space – a space in a building for living, sleeping, eating or cooking. Bathrooms, toilet rooms, closets, halls, storage, or utility spaces are not considered habitable spaces.

BATHROOM DESIGN

IRC 307 Toilet, Bath and Shower Spacing

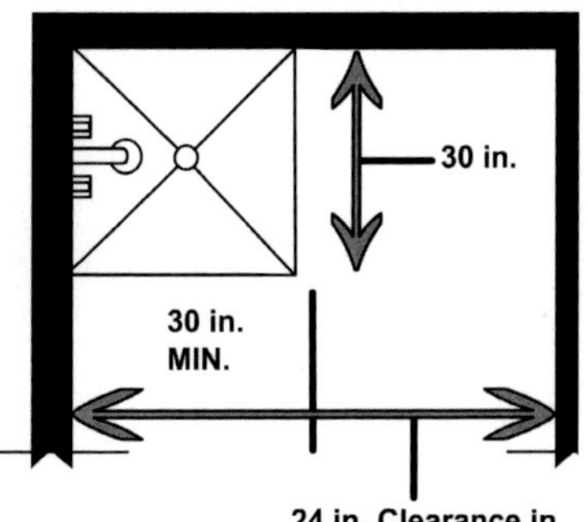

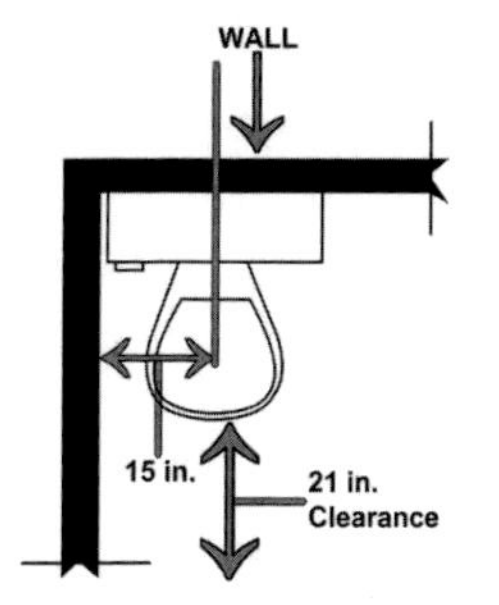

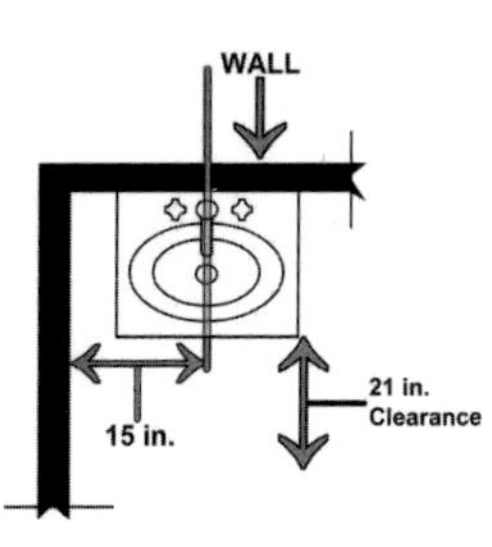

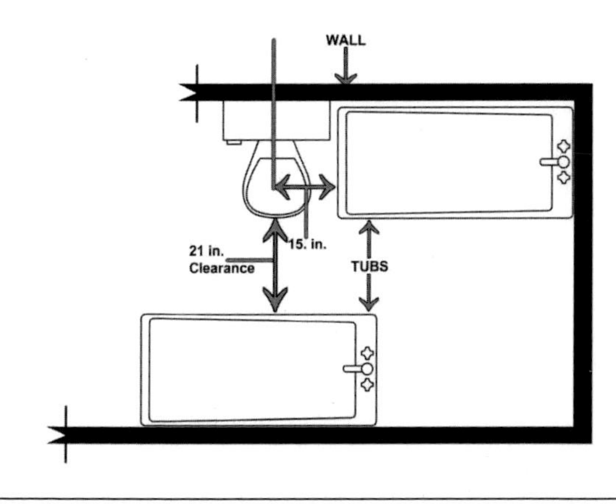

MEANS OF EGRESS

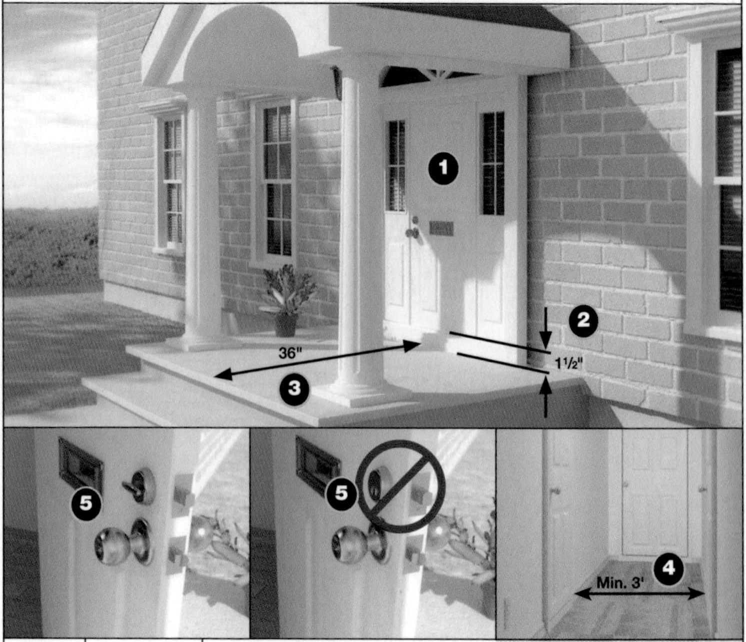

No.	Code	Description
1	R311.2	At least one egress side-hinged door required and shall provide a clear width of not less than 32" (between face of door and stop) and a clear height of not less than 78" (top of threshold to bottom of stop).
2	R311.3.1	Landing can not be more than 1.5" lower than the top of the threshold. EXCEPTION: Can be 7¾" if door swings in.
3	R311.3	Slope of the Landing can be no greater than 2% and landing must be at least as wide as the door it serves and 3' in the direction of travel.
4	R311.6	Minimum width of hallways is 3'.
5	R311.2	The one required egress door must be readily opened, without the use of a key, from the inside.

YOU SHOULD KNOW: R311.3.2

- **Exceptions:** Only applies to exterior doors that are not the required egress door.

EMERGENCY ESCAPE AND RESCUE OPENINGS

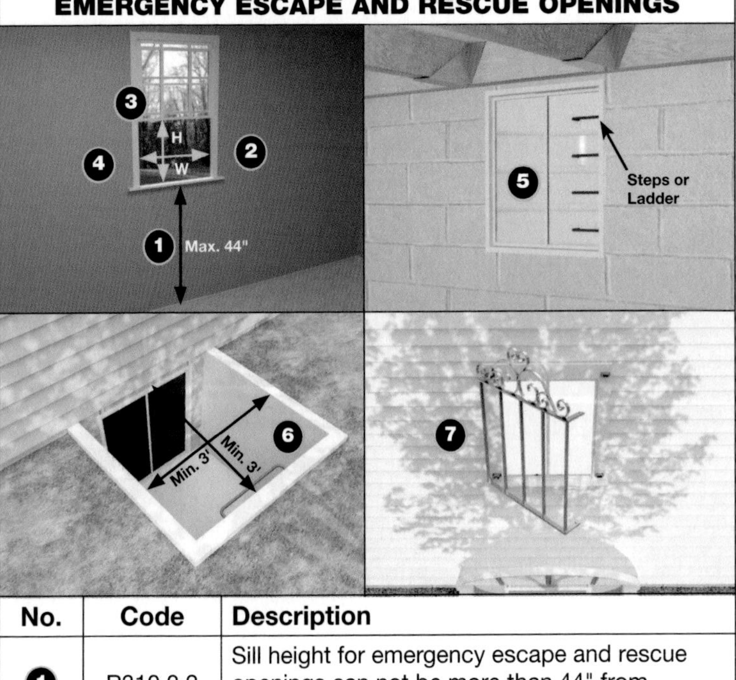

No.	Code	Description
1	R310.2.2	Sill height for emergency escape and rescue openings can not be more than 44" from the floor.
2	R310.2.1	Net opening width must be at least 20".
3	R310.2.1	Net opening height must be at least 24".
4	R310.2.1	Net opening must be 5.0 square feet for grade level windows and 5.7 square feet for all other levels.
5	R310.2.3.1	Window wells with a depth of more than 44" must have permanent ladders/steps.
6	R310.2.3	Window wells must be at least 3' wide and 3' in horizontal depth and not less than 9 square feet in horizontal area. *(The area must allow the escape window to be fully opened.)*
7	R310.4	Bars, grilles, covers and screens must be releasable/removable from the inside without use of key, tool or force greater than that required to open a window.

SAFETY GLAZING

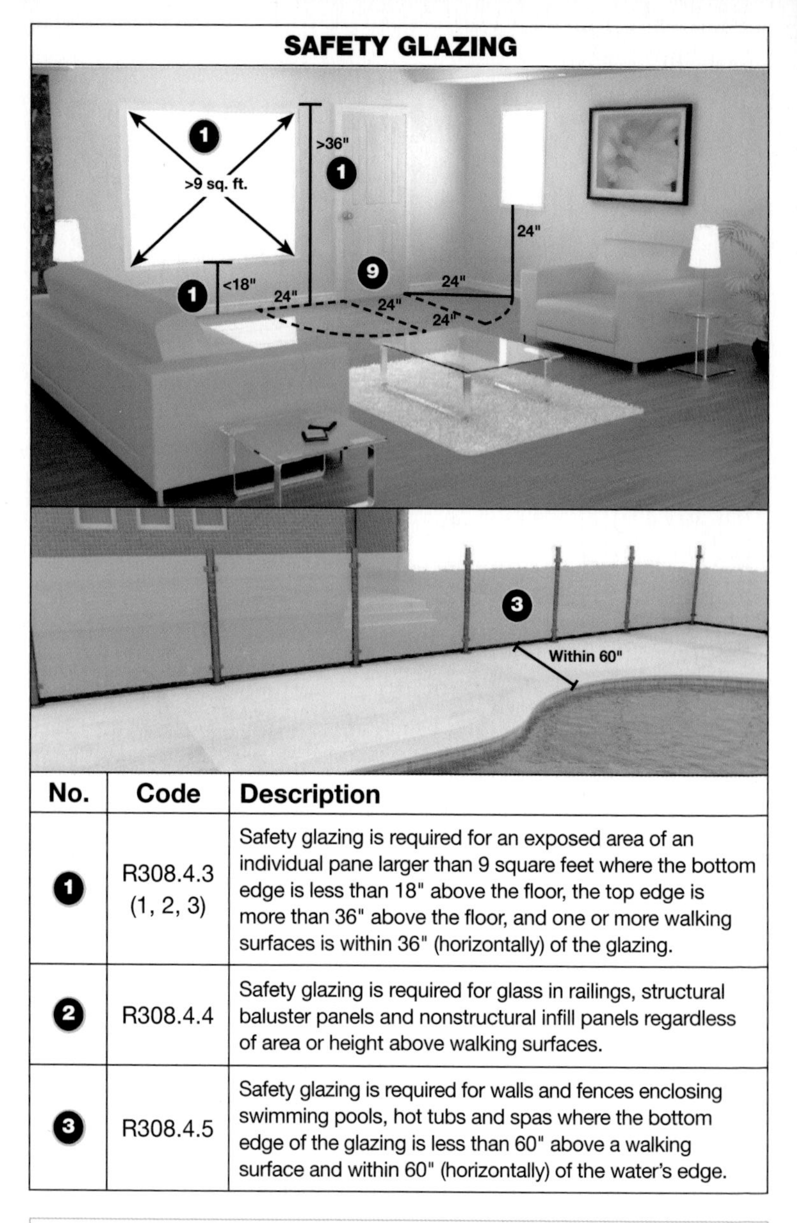

No.	Code	Description
1	R308.4.3 (1, 2, 3)	Safety glazing is required for an exposed area of an individual pane larger than 9 square feet where the bottom edge is less than 18" above the floor, the top edge is more than 36" above the floor, and one or more walking surfaces is within 36" (horizontally) of the glazing.
2	R308.4.4	Safety glazing is required for glass in railings, structural baluster panels and nonstructural infill panels regardless of area or height above walking surfaces.
3	R308.4.5	Safety glazing is required for walls and fences enclosing swimming pools, hot tubs and spas where the bottom edge of the glazing is less than 60" above a walking surface and within 60" (horizontally) of the water's edge.

YOU SHOULD KNOW: R308.4(3)

- Safety glazing required for glass adjacent to stairways, landings, and ramps within 36" horizontally of a walking surface when the exposed surface of the glass is less than 36" above the adjacent walking surface.

SAFETY GLAZING *(cont.)*

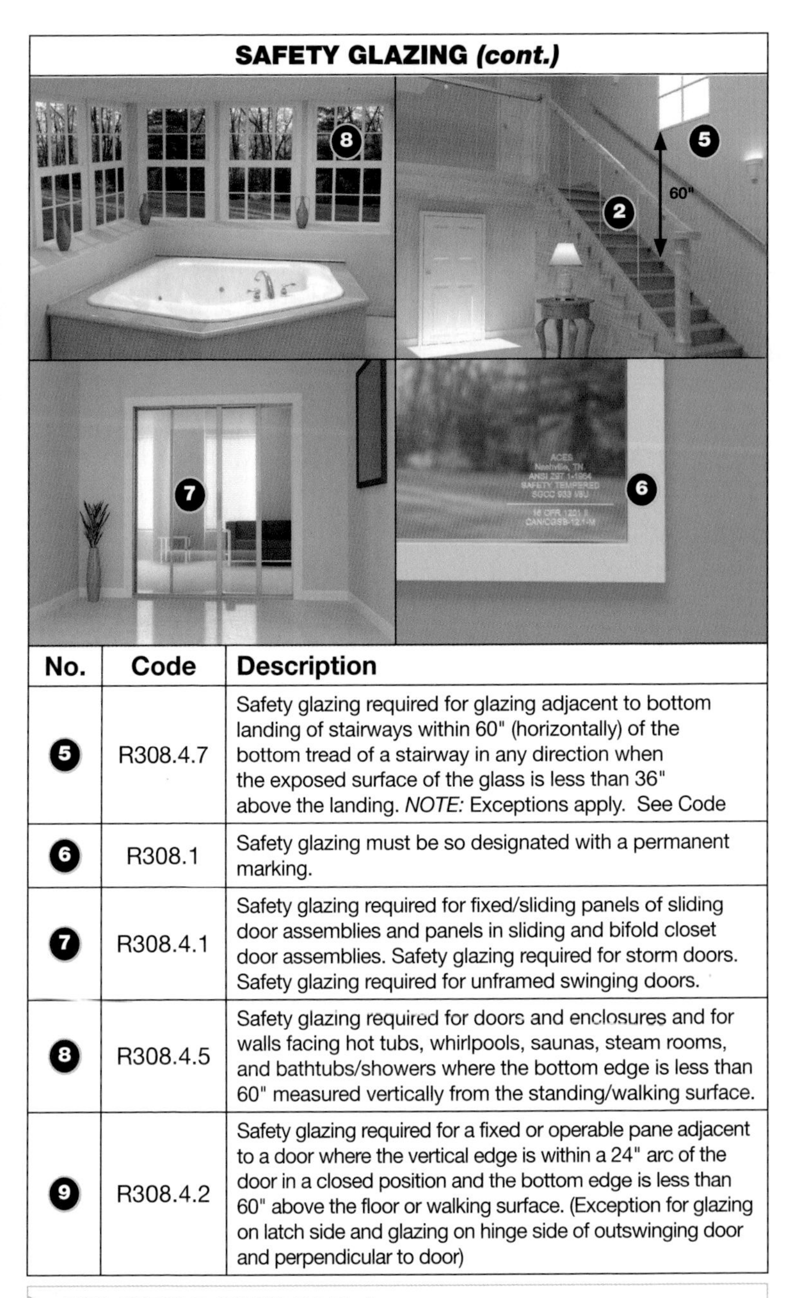

No.	Code	Description
5	R308.4.7	Safety glazing required for glazing adjacent to bottom landing of stairways within 60" (horizontally) of the bottom tread of a stairway in any direction when the exposed surface of the glass is less than 36" above the landing. *NOTE:* Exceptions apply. See Code
6	R308.1	Safety glazing must be so designated with a permanent marking.
7	R308.4.1	Safety glazing required for fixed/sliding panels of sliding door assemblies and panels in sliding and bifold closet door assemblies. Safety glazing required for storm doors. Safety glazing required for unframed swinging doors.
8	R308.4.5	Safety glazing required for doors and enclosures and for walls facing hot tubs, whirlpools, saunas, steam rooms, and bathtubs/showers where the bottom edge is less than 60" measured vertically from the standing/walking surface.
9	R308.4.2	Safety glazing required for a fixed or operable pane adjacent to a door where the vertical edge is within a 24" arc of the door in a closed position and the bottom edge is less than 60" above the floor or walking surface. (Exception for glazing on latch side and glazing on hinge side of outswinging door and perpendicular to door)

YOU SHOULD KNOW: R308.4

- Safety glazing required for doors (except jalousies).

GARAGES AND CARPORTS

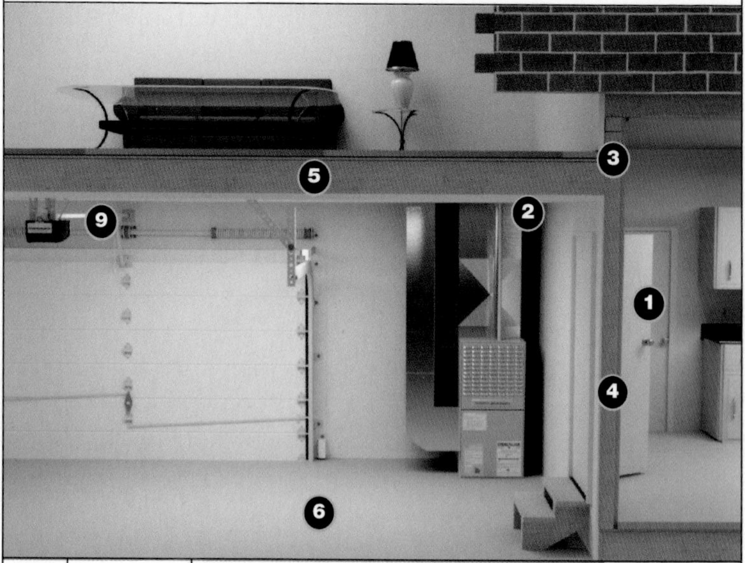

No.	Code	Description
1	R302.5.1	Doors separating a garage and residence must be: solid wood not less than 1 3/8" thick, solid or honeycomb core steel not less than 1 3/8" thick or 20 minute fire-rated.
2	R302.5.2	Ducts penetrating the dwelling from the garage must be at least 26 gauge, and have no openings into garage. All penetrations must be protected to prevent the free passage of fire.
3	R302.11 (2)	Fireblocking is required at all interconnections between concealed vertical and horizontal spaces *(soffits, drop ceilings and cove ceilings)*.
4	R302.6	Garage must be separated from the residence and its attic area by MIN 1/2" gypsum board applied to the garage side.

YOU SHOULD KNOW: R309.1

- ½" gypsum board must be installed on the interior side of the exterior wall of a detached garage located less than 3' from a dwelling unit.
- If a carport is not open on at least two sides, it is considered a garage.
- An opening from the garage to a room used for sleeping purposes is not allowed.

GARAGES AND CARPORTS *(cont.)*

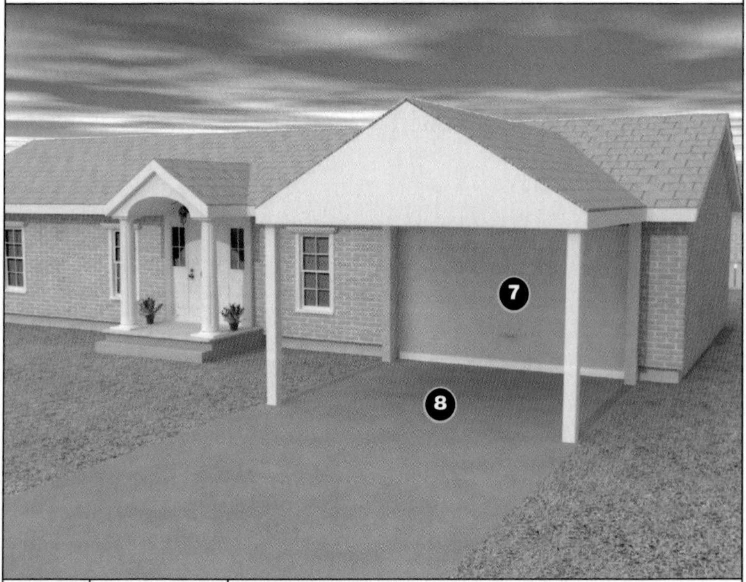

No.	Code	Description
5	R302.6	When a habitable room is above the garage, the ceiling in the garage must be covered with 5/8" Gypsum board (minimum) and support walls 1/2".
6	R309.1	Garage floors must be noncombustible and provide means of drainage *(either sloped to exterior or toward drain)*.
7	R309.2	A carport must be open on at least two sides. If not open on at least two sides, it will be considered a garage and must comply with the provisions of Section 309.
8	R309.2	Carport floor surface must be of approved non-combustible material *(asphalt is permitted)*.
	R309.2	The floor area of the carport used for parking any type of vehicle must be sloped for drainage to a drain or entry doorway.
9	R309.4	Door openers must comply with UL 325.

STAIRS

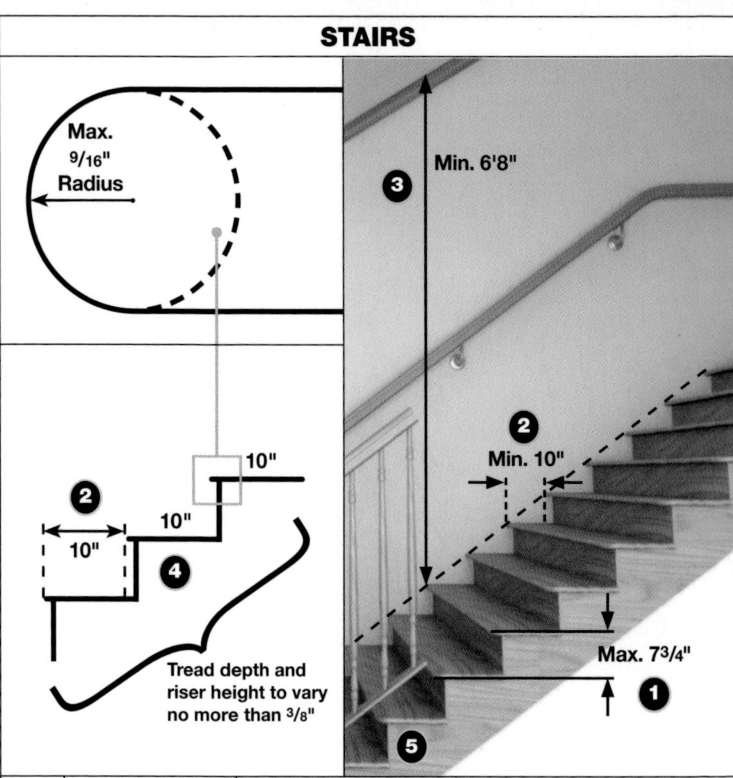

No.	Code	Description
1	R311.7.5.1	Maximum riser height is 7 3/4". *(Measured vertically between leading edges of the adjacent treads.)*
2	R311.7.5.2	Minimum tread depth is 10". *(Measured horizontally between the vertical planes of the foremost projection of adjacent treads and at a right angle to the tread's leading edge.)*
3	R311.7.2	Headroom must be at least 6'8" above plane of the tread nosings.
4	R311.7.5.1 & R311.7.5.2	Greatest tread depth or riser height within any flight of stairs must not exceed the smallest by more than 3/8".
5	R311.7.5.3	Nosing must be 3/4" to 1 1/4" for stairs with solid risers. Greatest projection must not exceed the smallest by more than 3/8".

YOU SHOULD KNOW: R311.5.5
Walking surface slopes not to exceed 2%.

STAIRS *(cont.)*

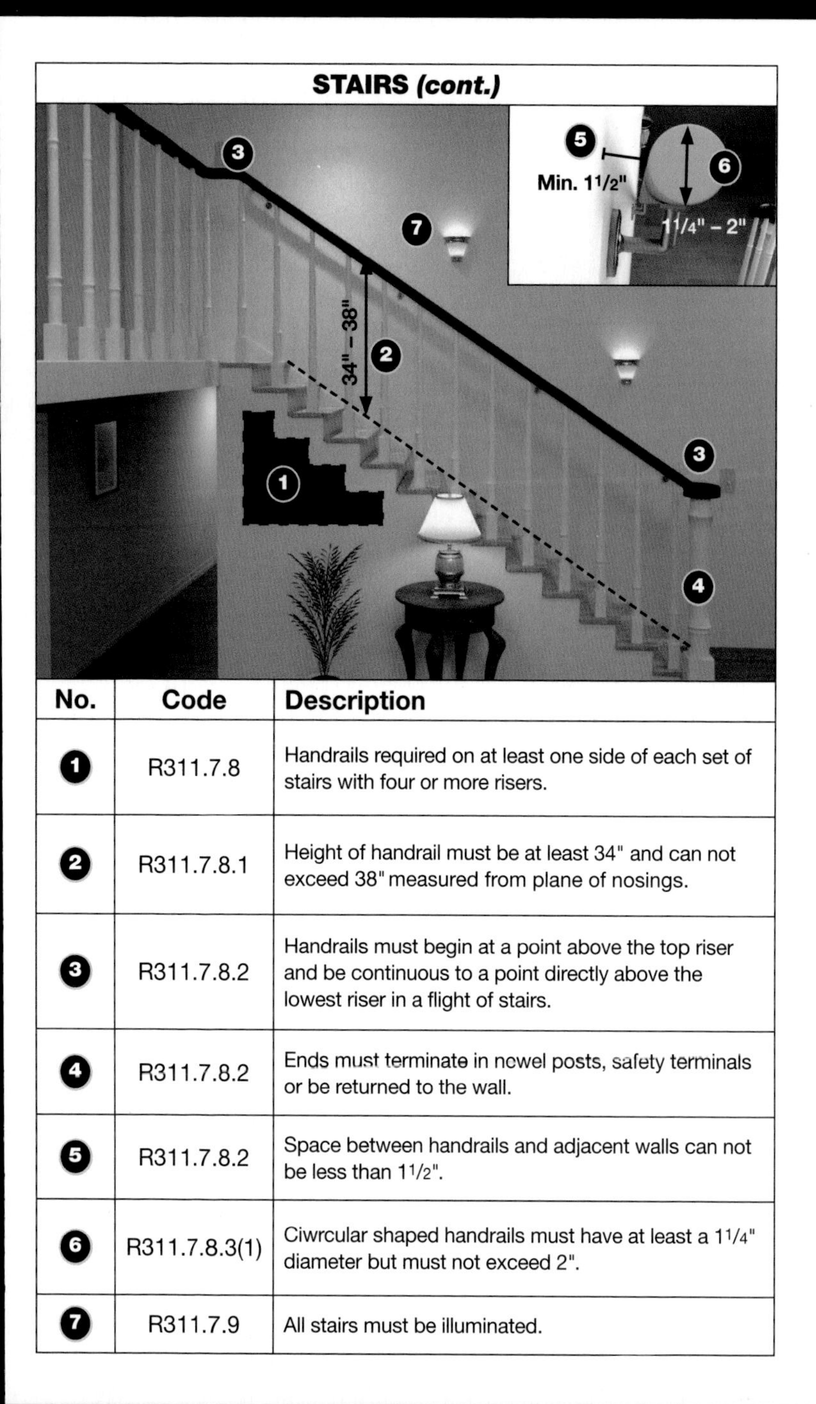

No.	Code	Description
1	R311.7.8	Handrails required on at least one side of each set of stairs with four or more risers.
2	R311.7.8.1	Height of handrail must be at least 34" and can not exceed 38" measured from plane of nosings.
3	R311.7.8.2	Handrails must begin at a point above the top riser and be continuous to a point directly above the lowest riser in a flight of stairs.
4	R311.7.8.2	Ends must terminate in newel posts, safety terminals or be returned to the wall.
5	R311.7.8.2	Space between handrails and adjacent walls can not be less than 1 1/2".
6	R311.7.8.3(1)	Ciwrcular shaped handrails must have at least a 1 1/4" diameter but must not exceed 2".
7	R311.7.9	All stairs must be illuminated.

STAIRS *(cont.)*

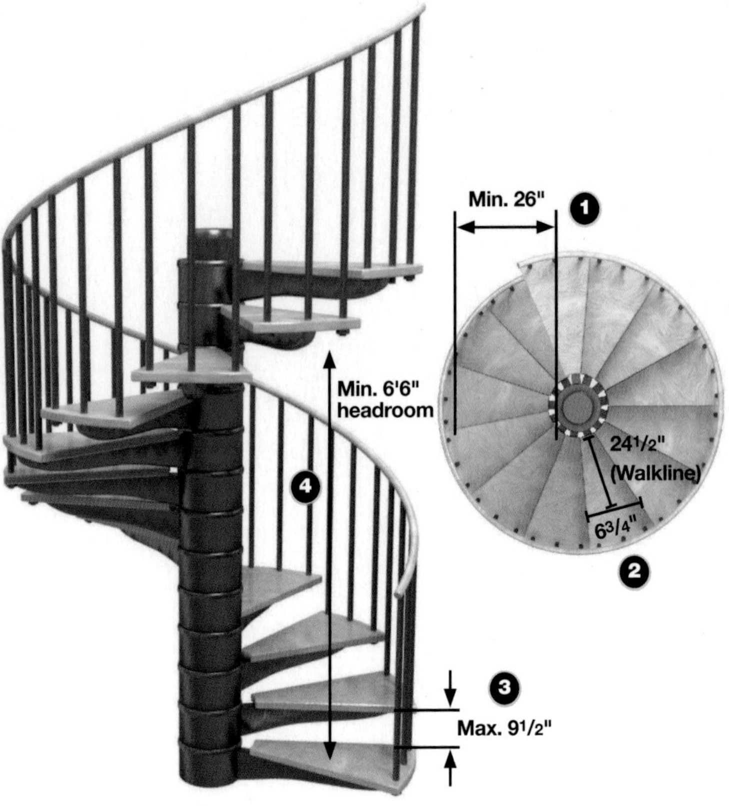

Min. 26"
Min. 6'6" headroom
$24^1/_2$" (Walkline)
$6^3/_4$"
Max. $9^1/_2$"

No.	Code	Description
1	R311.7.10.1	Minimum width of spiral stairs must be 26".
2	R311.7.10.1	Tread depth shall be minimum of 6¾" at the walkline.
3	R311.7.10.1	Riser height of spiral stairs can not be greater than $9^1/_2$".
4	R311.7.10.1	Headroom for spiral stairs is required to be 6'6".

GUARDS

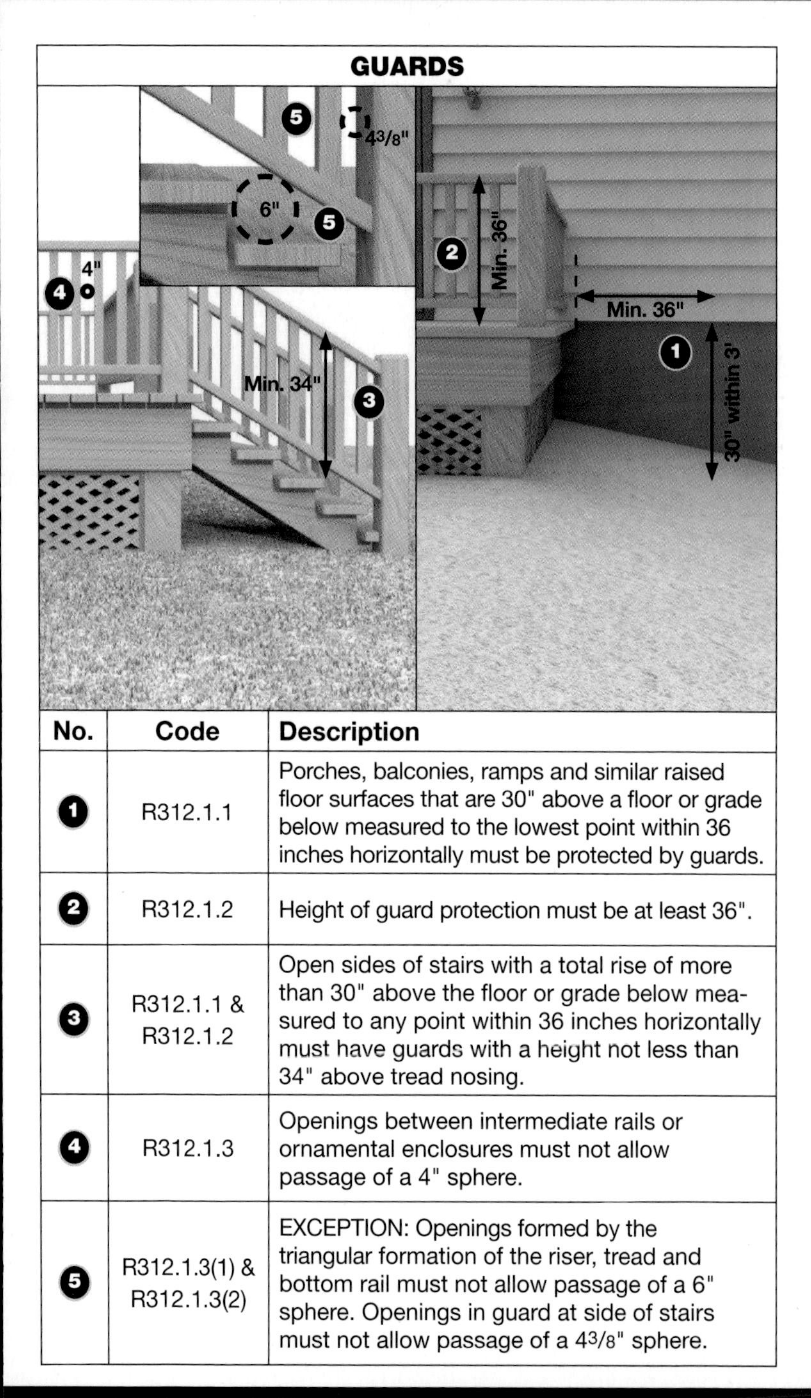

No.	Code	Description
1	R312.1.1	Porches, balconies, ramps and similar raised floor surfaces that are 30" above a floor or grade below measured to the lowest point within 36 inches horizontally must be protected by guards.
2	R312.1.2	Height of guard protection must be at least 36".
3	R312.1.1 & R312.1.2	Open sides of stairs with a total rise of more than 30" above the floor or grade below measured to any point within 36 inches horizontally must have guards with a height not less than 34" above tread nosing.
4	R312.1.3	Openings between intermediate rails or ornamental enclosures must not allow passage of a 4" sphere.
5	R312.1.3(1) & R312.1.3(2)	EXCEPTION: Openings formed by the triangular formation of the riser, tread and bottom rail must not allow passage of a 6" sphere. Openings in guard at side of stairs must not allow passage of a 4³/8" sphere.

SMOKE DETECTORS

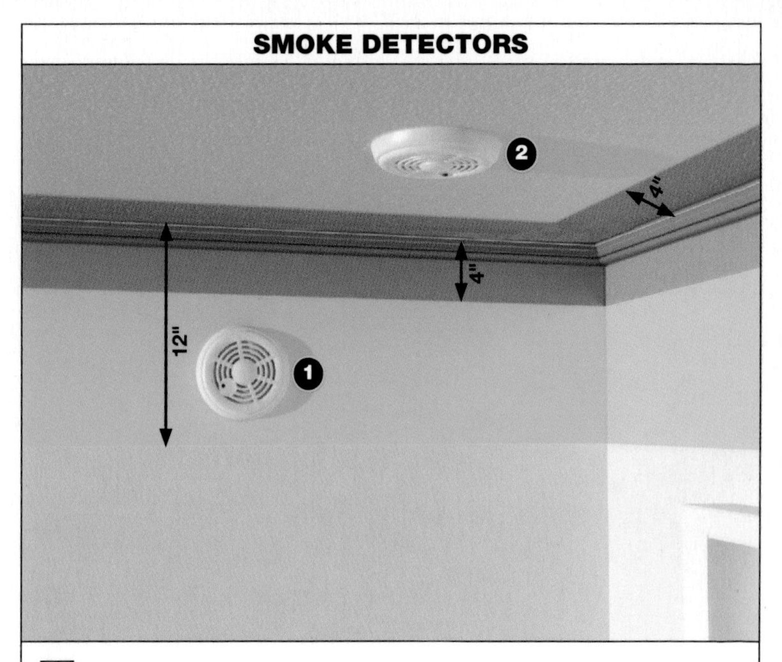

- Smoke detectors can be installed in this area
- Smoke detectors can NOT be installed in this area

No.	Code	Description
1	NFPA 72	Smoke detectors located on the wall must be located no closer than 4" from the ceiling and no further than 12" from the ceiling.
2	NFPA 72	Smoke detectors located on the ceiling must be located no closer than 4" from the wall.

YOU SHOULD KNOW:

- **2014 NEC:** 110.3(B) Requires all manufacturers instructions to be followed when installing listed equipment such as smoke detectors.
- NFPA 72 mandates smoke detector placement. Manufacturer's instructions provide these rules for smoke detector placement. Read instructions carefully.

SMOKE DETECTOR PLACEMENT
PEAKED CEILINGS
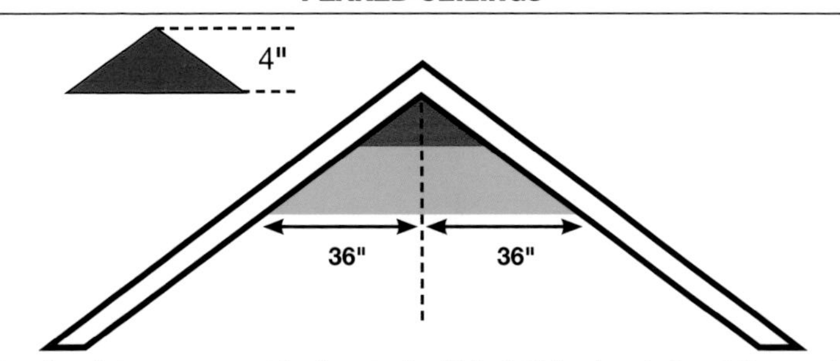 Smoke detectors must be located within 36" horizontally of the peak, but no closer than 4" vertically to the peak.
TRAY CEILINGS
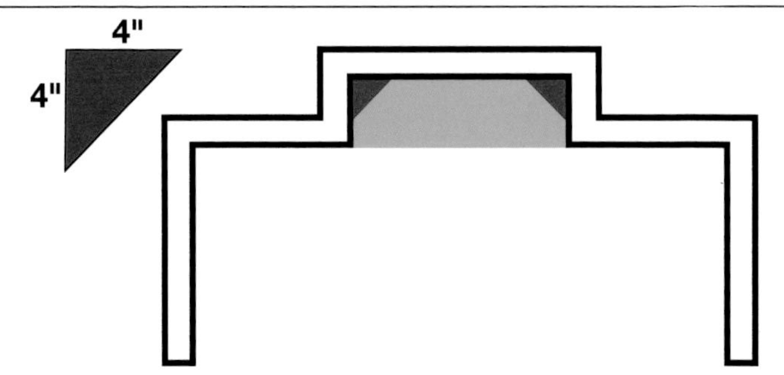 Smoke detectors must be installed on the highest point of the ceiling no closer than 4" from the sidewall or on the sidewall of the tray portion between 4" and 12" from the ceiling.
SLOPED CEILINGS
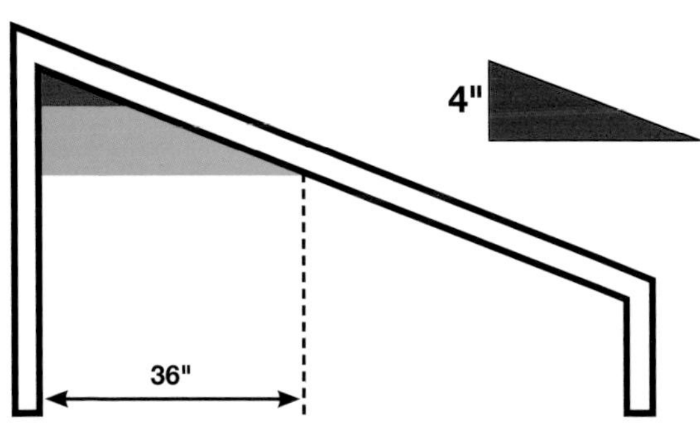 Smoke detectors must be installed within 36" of the high side of the ceiling, but not closer than 4" from the adjoining wall surface.

SMOKE DETECTOR LAYOUT

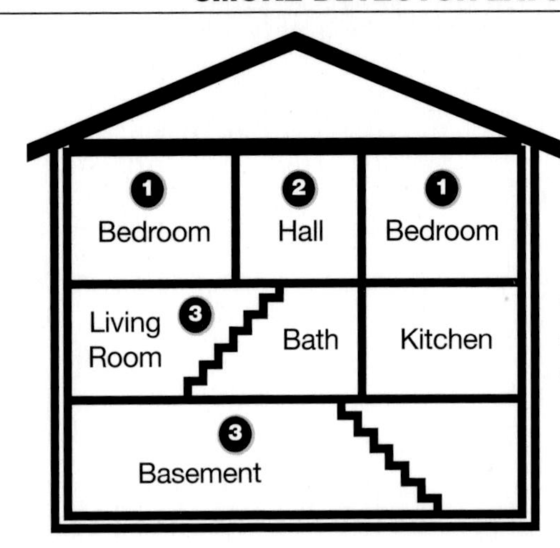

Typical section and floor layout of two different dwellings.

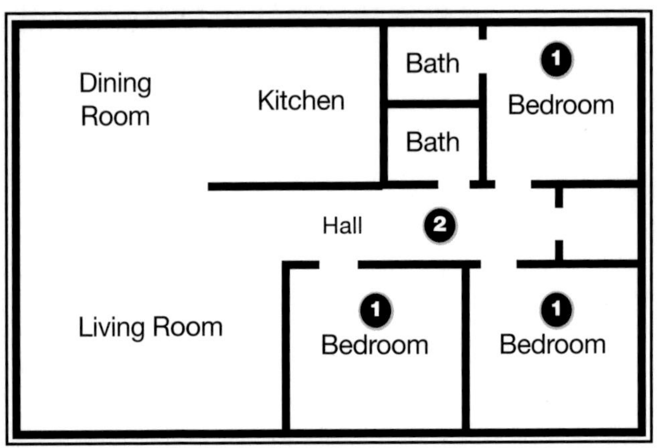

No.	Code	Description
1	R314.3(1)	Smoke detectors must be located in each sleeping room.
2	R314.3(2)	Smoke detectors must be located outside each sleeping area.
3	R314.3(3)	Smoke detectors must be located on each level of the dwelling unit including basements, but NOT including crawl spaces or uninhabitable attics.

SMOKE DETECTOR WIRING

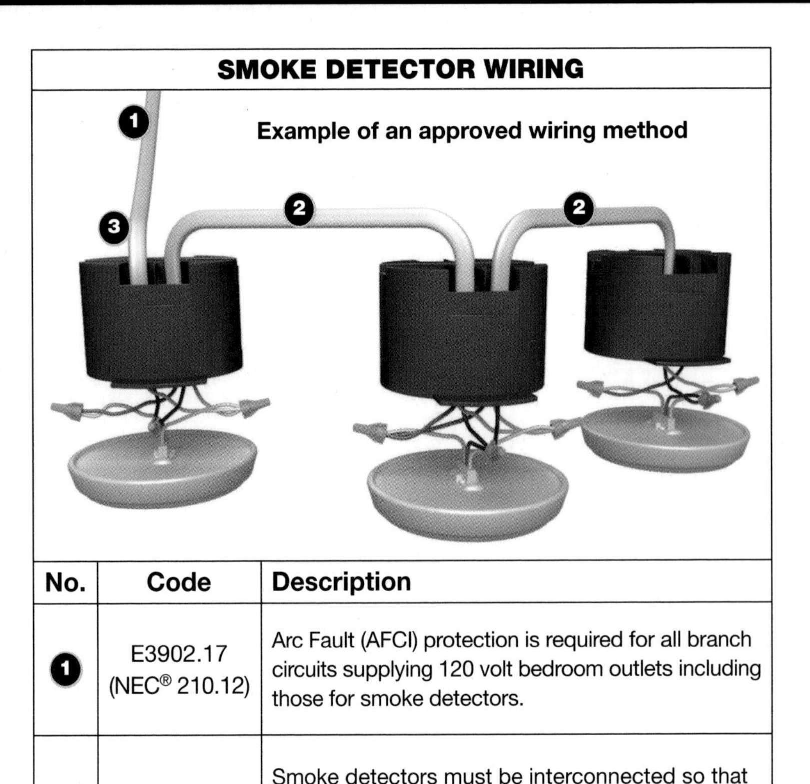

No.	Code	Description
1	E3902.17 (NEC® 210.12)	Arc Fault (AFCI) protection is required for all branch circuits supplying 120 volt bedroom outlets including those for smoke detectors.
2	R314.3	Smoke detectors must be interconnected so that activation of one alarm will activate alarms on all smoke detectors.
3	R314.4	Smoke detectors must receive their primary power from the building wiring and be provided with battery backup. Wiring must be permanent and without a disconnecting switch other than a circuit breaker.

YOU SHOULD KNOW:

- In a typical installation as shown, the power supply to the first smoke detector is a 2 conductor with ground NM cable. Power and interconnection to each successive smoke detector is made with a 3 conductor with ground NM cable.
- When alterations, repair or additions require a permit, or when one or more sleeping rooms are added to an existing dwelling, smoke detectors are required as per new construction. *(Exceptions may apply.)*

CRAWLSPACE VENTILATION

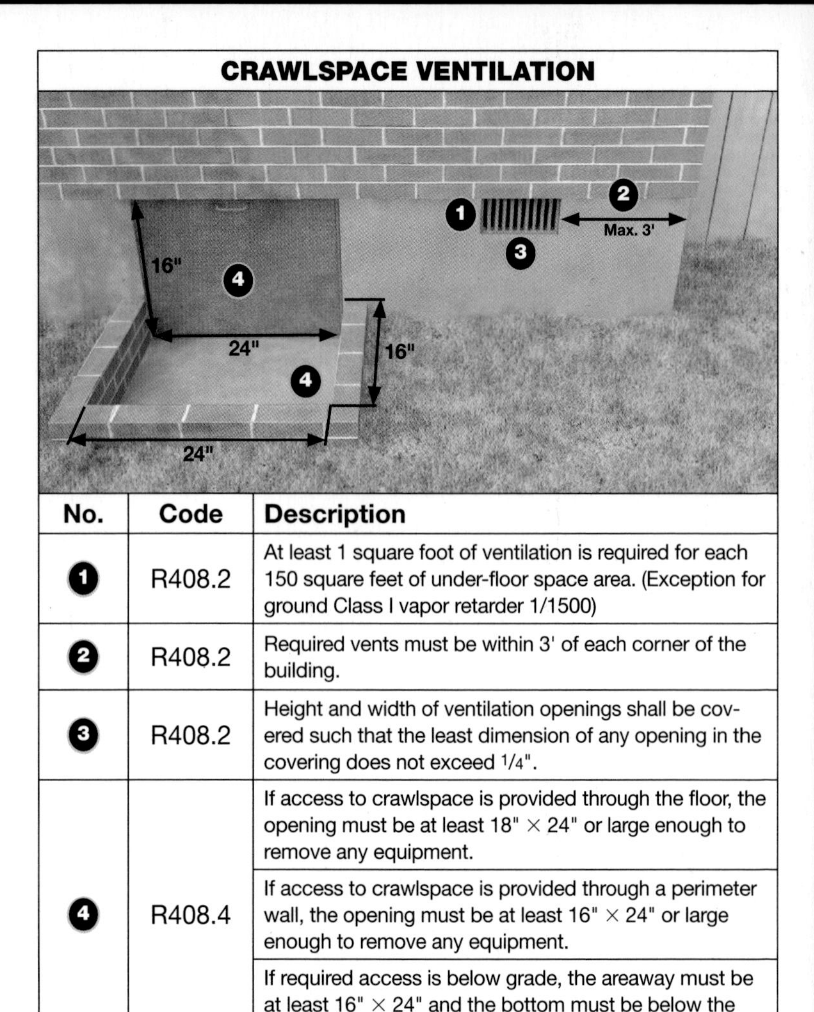

No.	Code	Description
1	R408.2	At least 1 square foot of ventilation is required for each 150 square feet of under-floor space area. (Exception for ground Class I vapor retarder 1/1500)
2	R408.2	Required vents must be within 3' of each corner of the building.
3	R408.2	Height and width of ventilation openings shall be covered such that the least dimension of any opening in the covering does not exceed 1/4".
4	R408.4	If access to crawlspace is provided through the floor, the opening must be at least 18" × 24" or large enough to remove any equipment.
		If access to crawlspace is provided through a perimeter wall, the opening must be at least 16" × 24" or large enough to remove any equipment.
		If required access is below grade, the areaway must be at least 16" × 24" and the bottom must be below the threshold of the opening.

YOU SHOULD KNOW:

- Vents are not required if:
  1. Exposed earth is covered with a continuous vapor retarder where joints lap by 6" and are sealed/taped. Edges must extend at least 6" up the stem wall and be attached. **R408.3**
  2. Perimeter walls are insulated and mechanical exhaust is operated continuously, or perimeter walls are insulated and conditioned air supply is provided or under-floor space is being used as a plenum. **R408.3**
- Perimeter wall access can not be located under a door to the residence. **R408.4**
- Crawlspace must be free of vegetation and all construction debris. **R408.5**

SOIL

IRC TABLE R401.4.1 – PRESUMPTIVE LOAD–BEARING VALUES OF FOUNDATION MATERIALS

Class of Material	Load–Bearing Pressure (PSF)
Crystalline bedrock	12,000
Sedimentary and foliated rock	4,000
Sandy gravel and/or gravel (GW and GP)	3,000
Sand, silty sand, clayey sand, silty gravel, and clayey gravel (SW, SP, SM, SC, GM, and GC)	2,000
Clay, sandy clay, silty clay, clayey silt, silt, and sandy silt (CL, ML, MH, and CH)	1,500

It is important to understand the type of soil you are working with as you plan for the building foundation. The building loads, both dead and live, will be transferred to the foundation footings. Consult with the local building authority to understand the soil type or test requirements for your area.

You may be required to conduct a sieve test which is a method to determine soil particle grain size. A sieve is a screen used to filter soil. It will allow particles smaller than its opening to fall through and retains larger particles. Sieve sizes are designated by the screen opening size. A 3" sieve has a screen with 3" square openings. If a soil sample passes the 3" sieve but fails the #4 sieve, the larger particle size is less than 3" and the smallest size is larger than 1/16". This soil would be classified as gravel.

Soils that pass the #4 sieve but fail on the #200 will be considered sand. Sands can further be categorized as coarse or fine sands.

Gradation is the distribution of different size groups within a soil sample. Well-graded soil has all sizes present from #4 to #200.

Poorly graded soil may be uniform (sieve sizes are all the same)

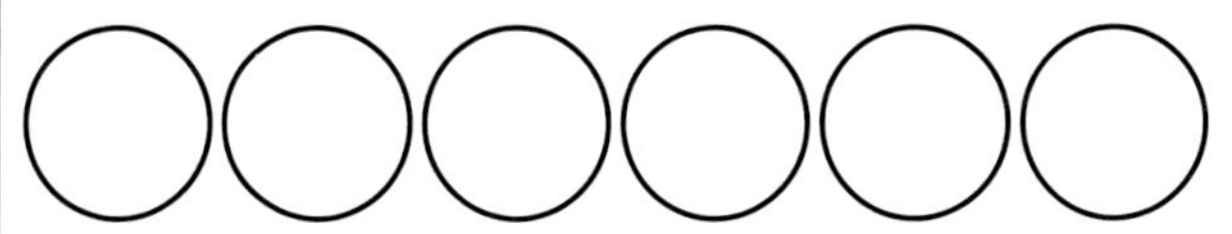

or can be gapped (have sieve sizes missing)

MIXING CONCRETE

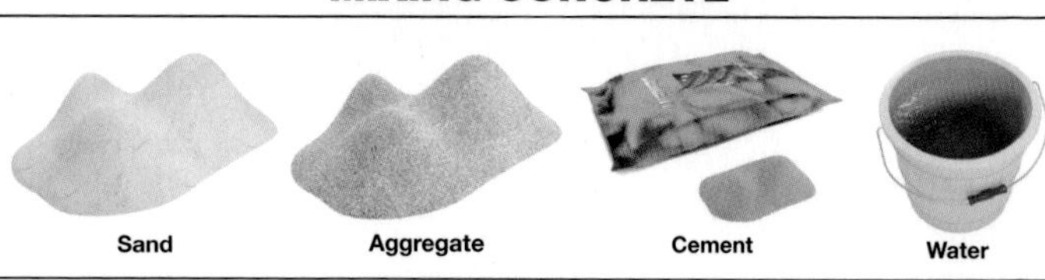

Concrete is formed by mixing cement, sand, water and aggregate. It is often necessary to modify the concrete mixture to achieve more durable concrete or make it more workable. Admixtures are commonly added for this purpose. Anything other than one of the four basic ingredients used to mix concrete is considered an admixture and include: accelerators, bonding admixtures, pumping aids, retarders, super plasticizers and water reducers to name a few.

TYPES OF CEMENT

Type	Purpose
I	General-purpose used most often.
II	Typically used for large footings or foundations where reduced heat of hydration is necessary or where moderate sulfate resistance is desired.
III	Used when high, early strength is required. This type of cement reduces the curing time from 28 days to 7 days.
IV	Low heat-of-hydration cement used for large concrete pours *(very large concrete projects)*.
V	Used in areas where sulfate conditions exist in the soil or where groundwater is found.

One of the most important components of concrete is cement. Cement is used to bind the water, sand and aggregate to form concrete. It has the most effect on the outcome of the mixture.

TYPES OF ADMIXTURES

Type	Purpose	Materials
Accelerator	Reduce the required time for setting and curing	Calcium chloride
Bonding Admixture	Increase bond strength	Acrylics, butadiene-styrene copolymers, polyvinyl chloride, and polyvinyl acetate
Pumping Aids	Allow easy pumping	Hydrated lime, organic and synthetic polymers, organic flocculents etc.
Retarders	Slows the set time	Sugars, borax, lignin, tartaric acid and salts
Super plasticizers	Increase workability of concrete without adding water	Sulfonated melamine or naphthalene, lignosulfonates, polycarboxylates etc.

COMPRESSIVE STRENGTH

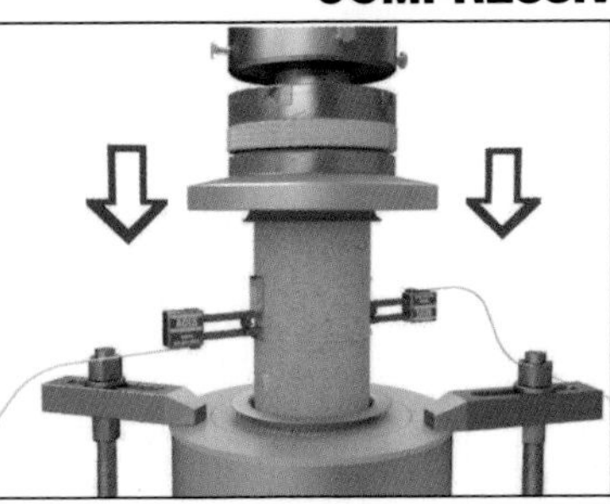

Images indicate a compressive strength test performed in a lab.

Compressive strength of concrete is measured in terms of PSI (pounds per square inch). This is the amount of pressure that the concrete will hold up to as the result from a force of a particular number of pounds applied to an area of one square inch.

Concrete with a 2500 PSI, will withstand a pressure equal to that of 2500 pounds exerted per square inch of the concrete. Application and weathering potential dictate the required strength.

TABLE 402.2 – MINIMUM SPECIFIED COMPRESSIVE STRENGTH OF CONCRETE AT 28 DAYS

Type or Location of Concrete Construction	Weathering Potential		
	Negligible	Moderate	Severe
Basement walls, foundations and other concrete not exposed to the weather	2,500	2,500	2,500
Basement slabs and interior slabs on grade, except garage floor slab.	2,500	2,500	2,500
Basement walls, foundation walls, and other vertical concrete work exposed to weather	2,500	3,000(a)	3,000(a)
Porches, carport slabs, and steps exposed to the weather, and garage floor slabs	2,500	3,000 (a)(b)(c)	3,500 (a)(b)(c)

(a) Concrete shall be air entrained. Total air content shall not be less than 5% or more than 7%.
(b) See section 402.2 for maximum cementious materials content.
(c) For garage floors with a steel troweled finish, reduction of the total air content to not less than 3% is permitted if the specified strangth of the concrete is increased to not less than 4000 psi.

CONCRETE SLUMP TEST – ASTM C143

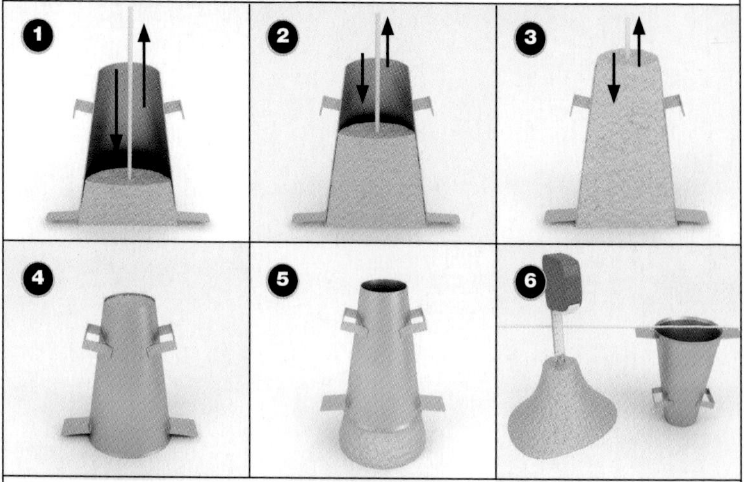

Testing the slump of the concrete as it is delivered to your job site is the only way you can be sure the concrete has been properly mixed. The slump test will also verify consistency of fresh concrete from one delivery to the next.

Slump test is not a typical requirement in residential construction under the IRC except for ICF walls or as required by an engineer.

It is important that the slump test be made within 5 minutes of taking the sample. Take two or more samples from the middle of the mixer discharge at regularly spaced intervals. Do not take samples for performing the test from the beginning or end of the discharge.

No.	Description
1	Hold the cone firmly in place, fill concrete to 1/3 of capacity and compact it by rodding 25 times with a round, straight steel rod. (5/8" x 24")
2	Continuing to hold the cone in place, add additional concrete to 2/3 capacity and repeat the rodding procedure.
3	Finally, fill the cone to overflowing capacity and repeat the rodding procedure. Be sure to distribute the strokes evenly.
4	Remove the excess concrete from the top of the cone so that it is level and full.
5	Within seconds of completing step 4, carefully remove the cone from the concrete. After the concrete loses the form of the cone, it will begin to slump downward.
6	Place the steel rod across the inverted cone. Measure from the bottom of the steel rod to the center of the top of the concrete. This measurement represents the slump.

REINFORCING

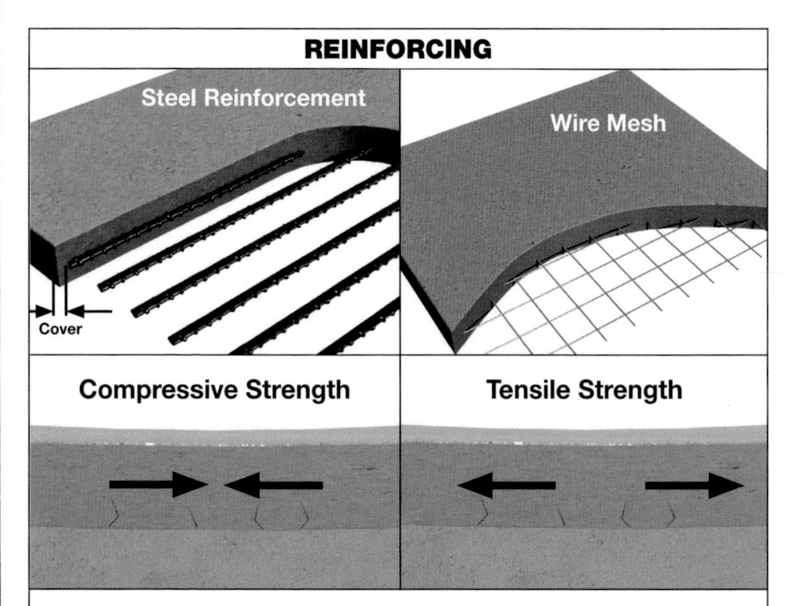

By placing rebar in concrete, the tensile strength is greatly improved. Concrete, without reinforcement, can resist crushing loads but its ability to resist being pulled apart (tensile strength) is weak. By combining reinforcement in concrete, it will become strong against tensile and compressive resistance.

The proper "Cover" (the distance from the outside of the bar to the surface of the concrete) should be maintained to protect the rebar from corrosion or damage. If not specified by engineer's drawings, the following minimum covers should be maintained.

MINIMUM COVER DISTANCE

Application	Size	Cover
Footing *(concrete cast against and permanently exposed to earth)*	—	3"
Basement Walls *(exposed to weather or earth)*	#6 – #18 rebar	2"
	#16, W31/D31 wire or smaller	1 1/2"
Slabs, Walls and Joists *(not exposed to weather or in contact with earth)*	#14 and #18 rebar	1/2"
	#11 rebar and smaller	3/4"
	Beams and columns	1 1/2"

FOOTINGS

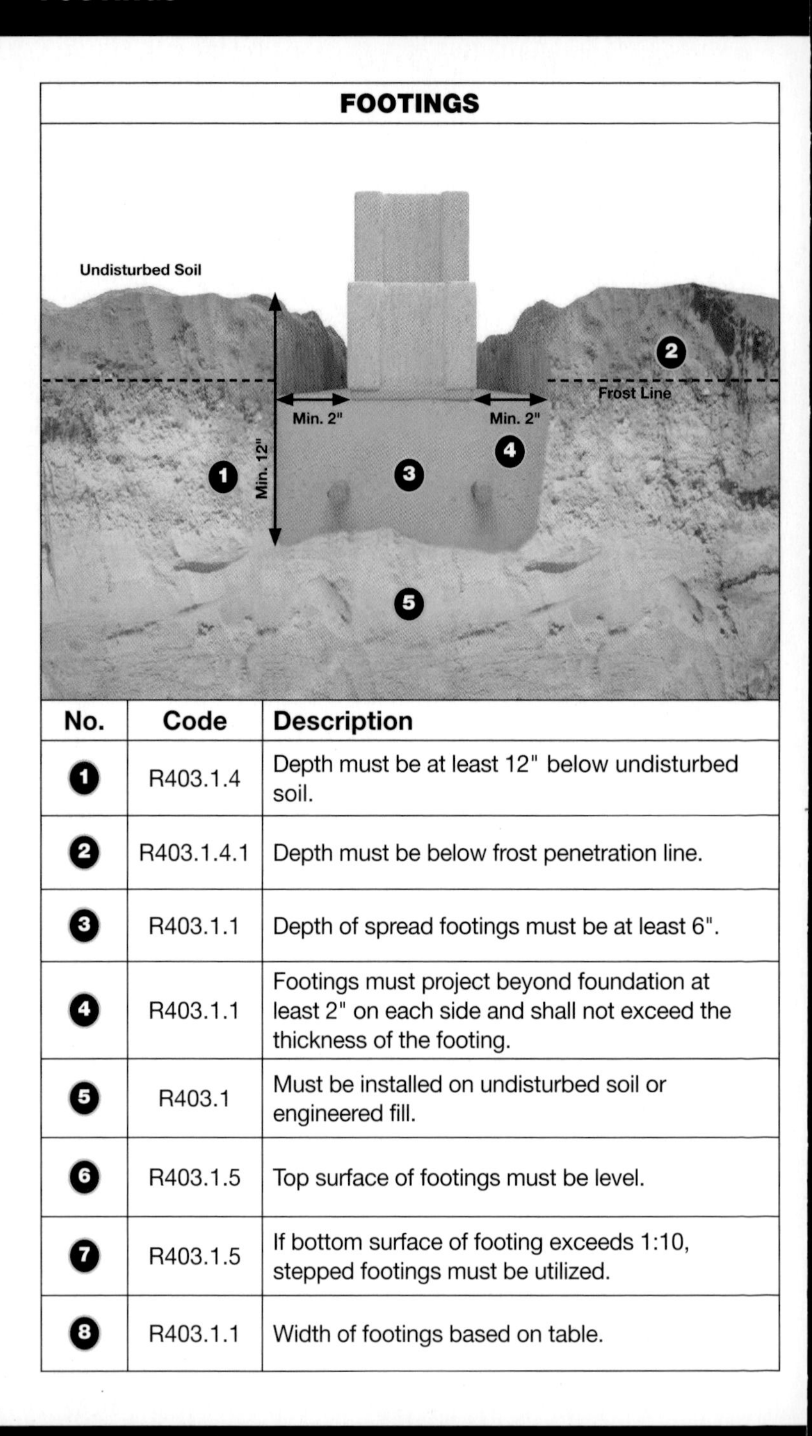

No.	Code	Description
1	R403.1.4	Depth must be at least 12" below undisturbed soil.
2	R403.1.4.1	Depth must be below frost penetration line.
3	R403.1.1	Depth of spread footings must be at least 6".
4	R403.1.1	Footings must project beyond foundation at least 2" on each side and shall not exceed the thickness of the footing.
5	R403.1	Must be installed on undisturbed soil or engineered fill.
6	R403.1.5	Top surface of footings must be level.
7	R403.1.5	If bottom surface of footing exceeds 1:10, stepped footings must be utilized.
8	R403.1.1	Width of footings based on table.

DECAY RESISTANCE

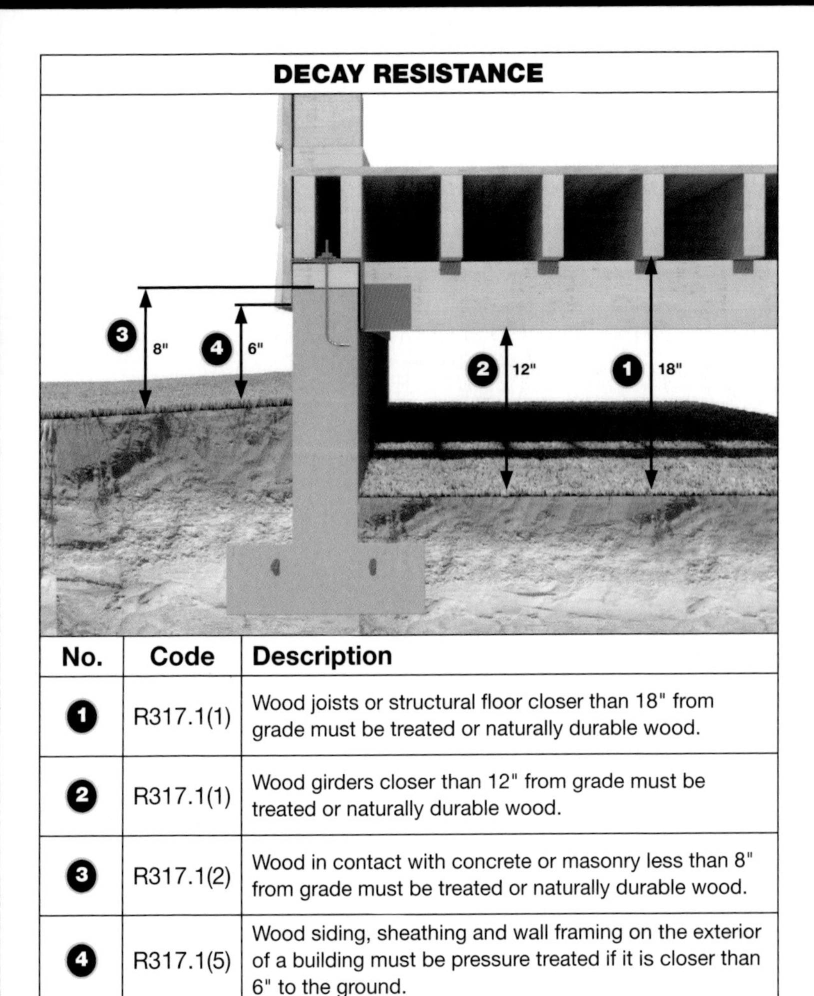

No.	Code	Description
1	R317.1(1)	Wood joists or structural floor closer than 18" from grade must be treated or naturally durable wood.
2	R317.1(1)	Wood girders closer than 12" from grade must be treated or naturally durable wood.
3	R317.1(2)	Wood in contact with concrete or masonry less than 8" from grade must be treated or naturally durable wood.
4	R317.1(5)	Wood siding, sheathing and wall framing on the exterior of a building must be pressure treated if it is closer than 6" to the ground.

YOU SHOULD KNOW: R317.1

- Members that support moisture-permeable floors or roofs that are exposed to the weather must be pressure treated unless separated by a moisture barrier.
- Wood furring strips or other wood framing members attached to the interior of exterior walls or concrete below grade must be pressure treated or naturally durable wood unless separated by a vapor barrier. (Pressure treated is required for ground contact and for wood embedded in concrete with ground contact.)
- Sills and Sleepers placed on concrete or masonry that is in contact with ground and not separated from slab by barrier must be treated.
- Ends of wood girders that enter exterior masonry or concrete walls must be protected against decay (Treated) if the clearance is less than 1/2".

SLAB ON GRADE

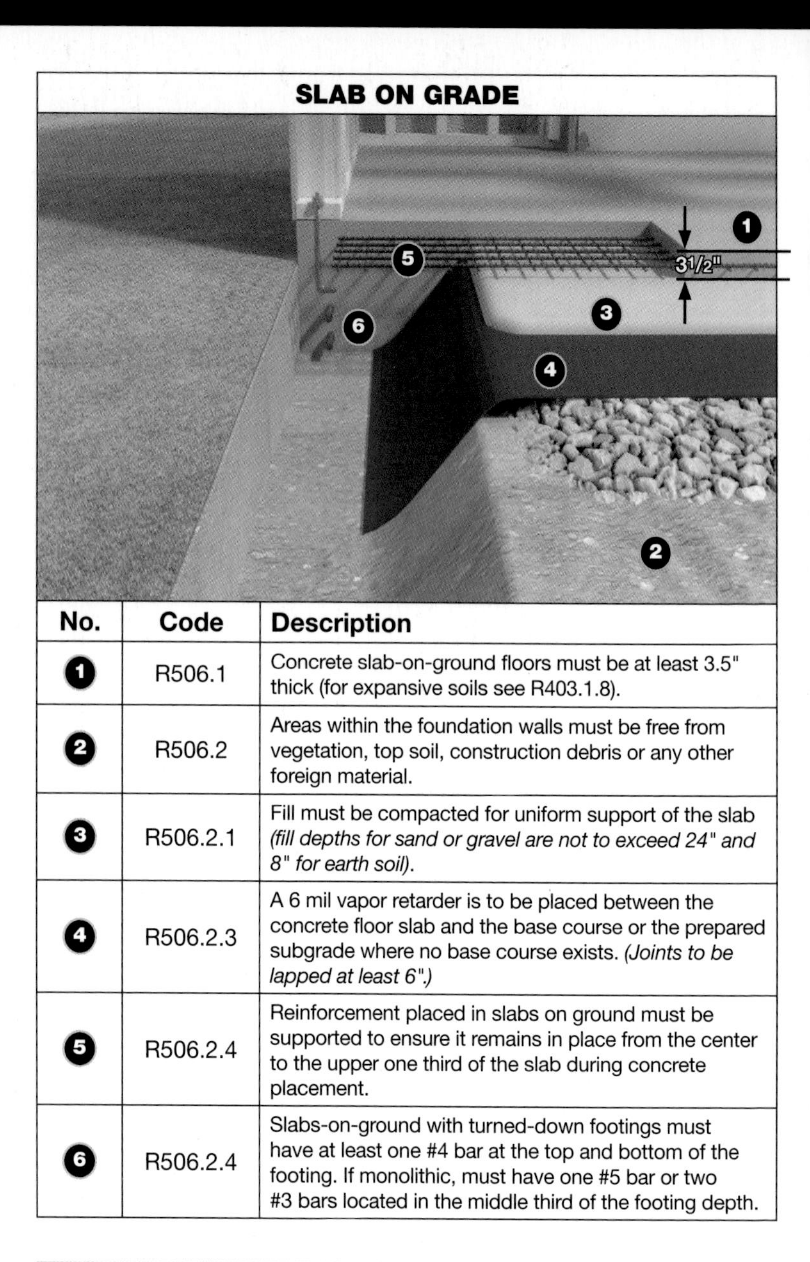

No.	Code	Description
1	R506.1	Concrete slab-on-ground floors must be at least 3.5" thick (for expansive soils see R403.1.8).
2	R506.2	Areas within the foundation walls must be free from vegetation, top soil, construction debris or any other foreign material.
3	R506.2.1	Fill must be compacted for uniform support of the slab *(fill depths for sand or gravel are not to exceed 24" and 8" for earth soil)*.
4	R506.2.3	A 6 mil vapor retarder is to be placed between the concrete floor slab and the base course or the prepared subgrade where no base course exists. *(Joints to be lapped at least 6".)*
5	R506.2.4	Reinforcement placed in slabs on ground must be supported to ensure it remains in place from the center to the upper one third of the slab during concrete placement.
6	R506.2.4	Slabs-on-ground with turned-down footings must have at least one #4 bar at the top and bottom of the footing. If monolithic, must have one #5 bar or two #3 bars located in the middle third of the footing depth.

YOU SHOULD KNOW: R506.2.2

- A base course of 4" thick clean graded sand, gravel, crushed stone or crushed blast-furnace slag passing a 2" sieve is to be placed on the prepared subgrade if the slab is below grade.

ANCHOR BOLT PLACEMENT

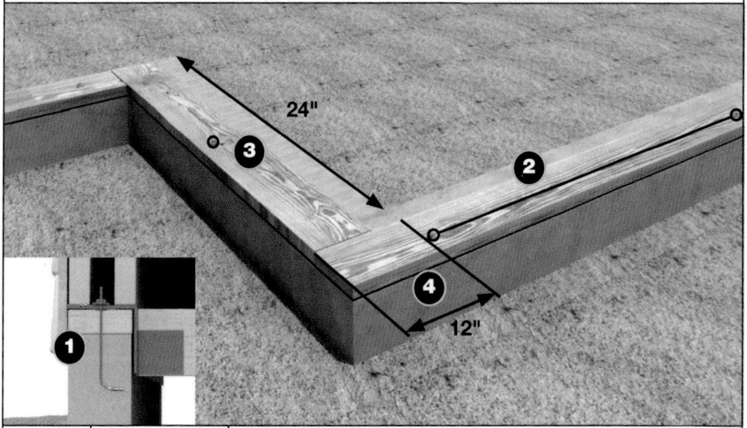

No.	Code	Description
1	R403.1.6	Anchor bolts must be at least 1/2" in diameter and must extend at least 7" into masonry or concrete.
2	R403.1.6	There must be at least 2 anchor bolts per plate section.
	R403.1.6	Anchor bolts must be spaced no more than 6' on center.
3	R403.1.6 Exc. #1	Walls 24" or shorter connecting an offset are allowed to be anchored with one bolt in the center third of the plate section and must be attached to the adjacent braced wall panels at corners.
4	R403.1.6	One anchor bolt must be located not more than 12" (or less than seven bolt diameters) from each end of the plate section.

YOU SHOULD KNOW: R403.1.6

- Walls 12" or shorter connecting offset braced wall panels can be connected to the foundation without anchor bolts.
- A nut and washer must be tightened on each bolt of the plate.
- Sill and sole plates must be protected against decay and termites. *(Pressure Treated)*

FOUNDATIONS

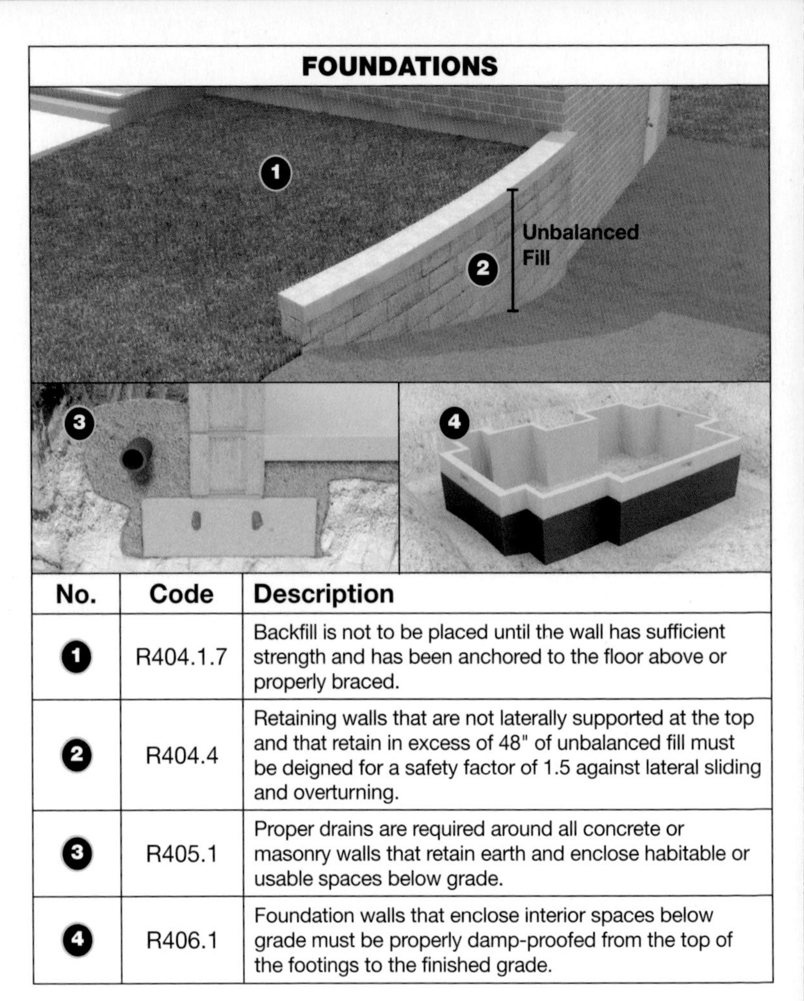

No.	Code	Description
1	R404.1.7	Backfill is not to be placed until the wall has sufficient strength and has been anchored to the floor above or properly braced.
2	R404.4	Retaining walls that are not laterally supported at the top and that retain in excess of 48" of unbalanced fill must be deigned for a safety factor of 1.5 against lateral sliding and overturning.
3	R405.1	Proper drains are required around all concrete or masonry walls that retain earth and enclose habitable or usable spaces below grade.
4	R406.1	Foundation walls that enclose interior spaces below grade must be properly damp-proofed from the top of the footings to the finished grade.

CUTTING AND NOTCHING

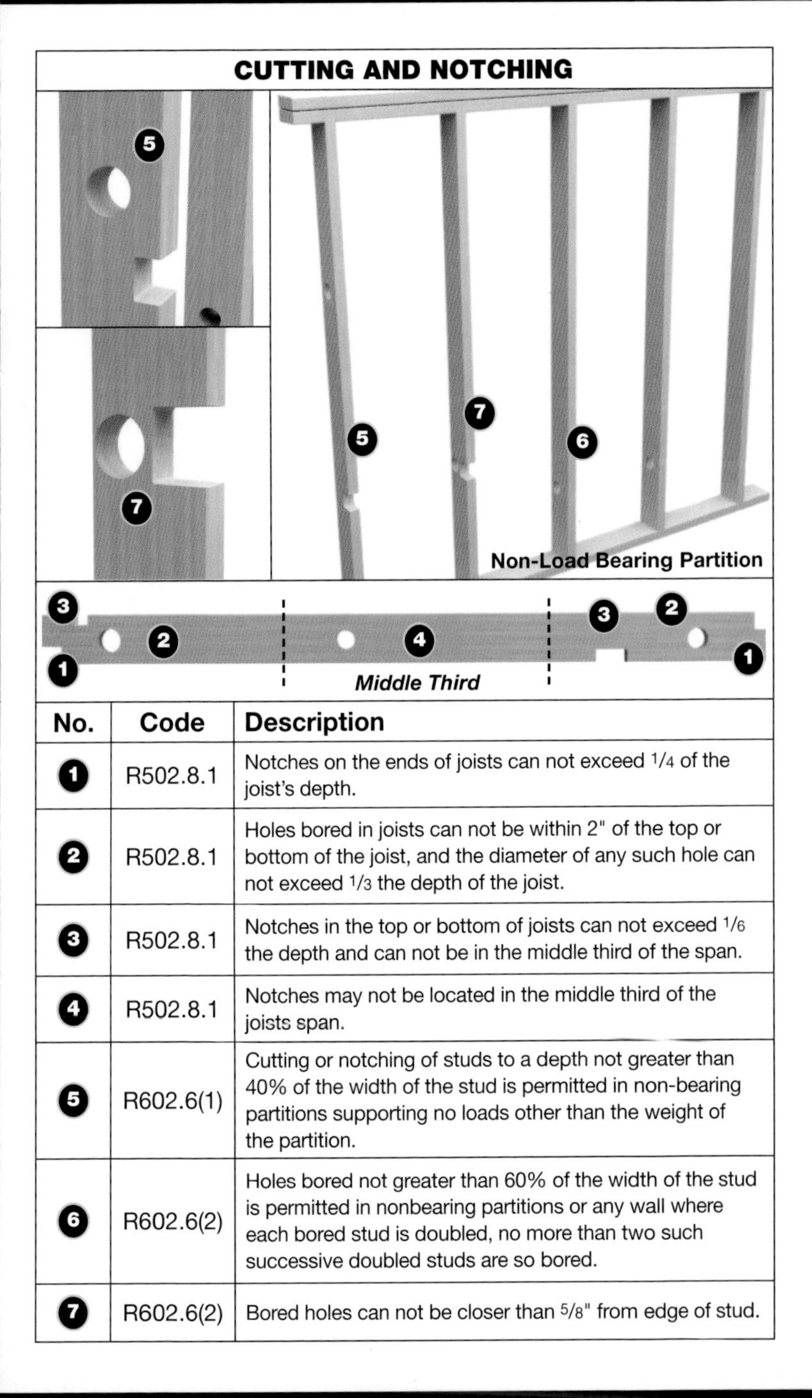

No.	Code	Description
1	R502.8.1	Notches on the ends of joists can not exceed 1/4 of the joist's depth.
2	R502.8.1	Holes bored in joists can not be within 2" of the top or bottom of the joist, and the diameter of any such hole can not exceed 1/3 the depth of the joist.
3	R502.8.1	Notches in the top or bottom of joists can not exceed 1/6 the depth and can not be in the middle third of the span.
4	R502.8.1	Notches may not be located in the middle third of the joists span.
5	R602.6(1)	Cutting or notching of studs to a depth not greater than 40% of the width of the stud is permitted in non-bearing partitions supporting no loads other than the weight of the partition.
6	R602.6(2)	Holes bored not greater than 60% of the width of the stud is permitted in nonbearing partitions or any wall where each bored stud is doubled, no more than two such successive doubled studs are so bored.
7	R602.6(2)	Bored holes can not be closer than 5/8" from edge of stud.

CUTTING AND NOTCHING *(cont.)*

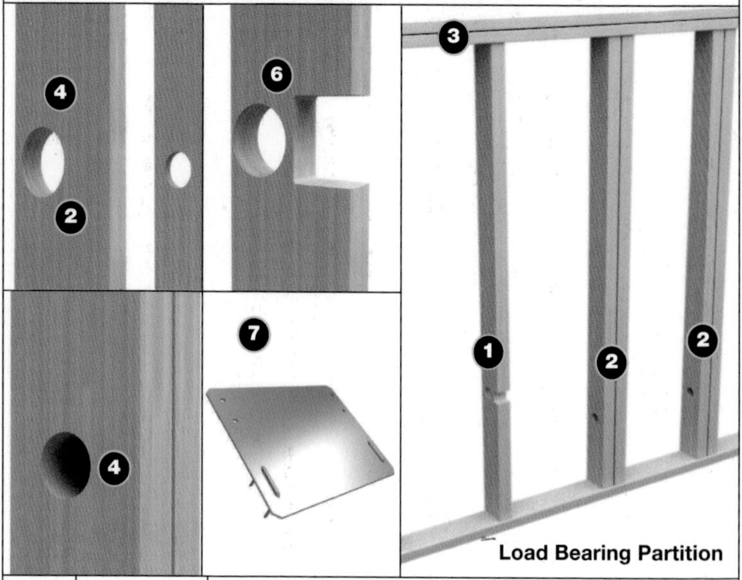

Load Bearing Partition

No.	Code	Description
1	R602.6(1)	In exterior walls and load bearing partitions, any wood stud is permitted to be cut or notched to a depth not exceeding 25% of its width.
2	R602.6(1)	In exterior walls and load bearing partitions, any bored hole can not be greater in diameter than 40% of the stud width.
3	R602.6.1 (8 IOD)	If top plate is cut, drilled, or notched by more than 50% of its width, a galvanized metal tie of not less than 0.054 inch thick and 1 1/2" wide must be fastened across and to the plate at each side of the opening with not less than eight 10d common nails at each side or equivalent.
4	R602.6(2)	Bored holes can not be closer than 5/8" from edge of stud.
5	R602.6(2)	Bored holes shall not be located in the same cross section of cut or notch in stud.
6	R602.6(2)	If hole is between 40% and 60% of the stud depth, then stud must be doubled and no more than two successive studs are doubled and so bored.
7	R2603.2.1	Steel shield plates must be installed where plumbing piping (other than cast-iron or galvanized steel) is installed through holes or notches in framing members less that 1¼" from the nearest edge of the member. The plate must cover the area where the member is notched or bored and extend at least 2" above sole plates and below top plates.

NAIL SIZES

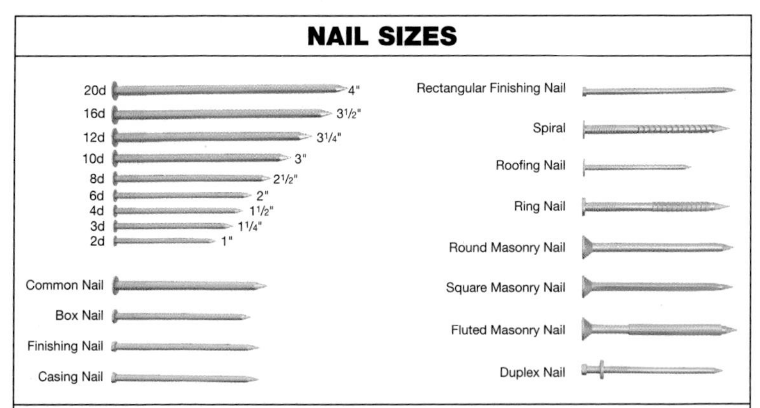

The nail sizes based on the term "penny" is believed to have first come into use in the early 1600's. The English monetary unit was the Pound Sterling which was divided into Shillings and Pence. The cost of 100 nails in Pence in the 1600's is how we refer to nail sizes today. For example, 100 nails that sold for 6 pence were called 6d nails. (6d is the abbreviation of 6 pence). 100 larger nails that sold for 12 pence are 12d nails and so on. Although the price of nails did not standardize the size, the constant established by such led to the standard sizes as we refer to them today.

TYPES OF NAILING

FACE NAILING

The nail is driven directly through the thickness of the member and into the component to which it is being attached.

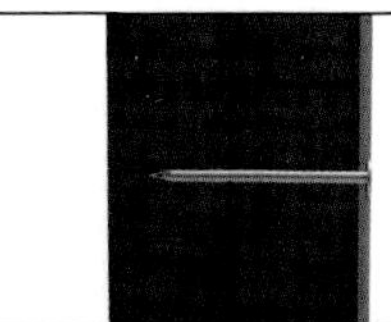

TOE NAILING

The nail is driven at an angle through the face (thickness) of the lumber and into the component to which it is being attached.

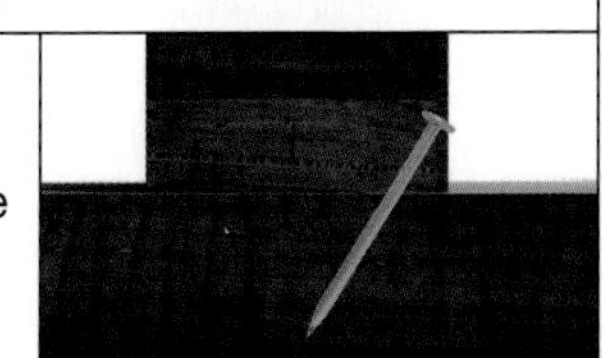

BLIND NAILING

Similar to toe nailing in that the nail is driven at an angle but different in that it will be concealed by the next board (component) to be attached.

JOISTS

No.	Code	Description
1	R502.6	Ends of each joist, beam or girder must bear at least 1.5" on wood or metal (3" on masonry or concrete).
2	R502.4	Joists under parallel bearing partitions must be of adequate size to support the load.
	R502.6.1	Joists framing from opposite sides over a bearing support must lap at least 3" and be nailed together with at least three 10d face nails.
3	R502.6.2	Joists framing into the side of a wood girder must be supported by approved anchors or on 2" × 2" ledger strips.
4	R502.7	Joists must be supported laterally at the ends by full-depth blocking not less than 2" nominal in thickness (or attached to a full-depth header, band or rim joist).
5	R502.7.1	Joists that EXCEED 2" × 12" nominal size must be supported laterally by solid blocking, diagonal bridging (wood or metal) or a continuous strip nailed across the bottom of joists perpendicular to joists at intervals not exceeding 8'.
6	R502.3.1 (1) & (2)	Floor joist spans should be based on Tables R502.3.1(1) & (2) *(See Tables on pages 63 and 64.)*

OPENINGS

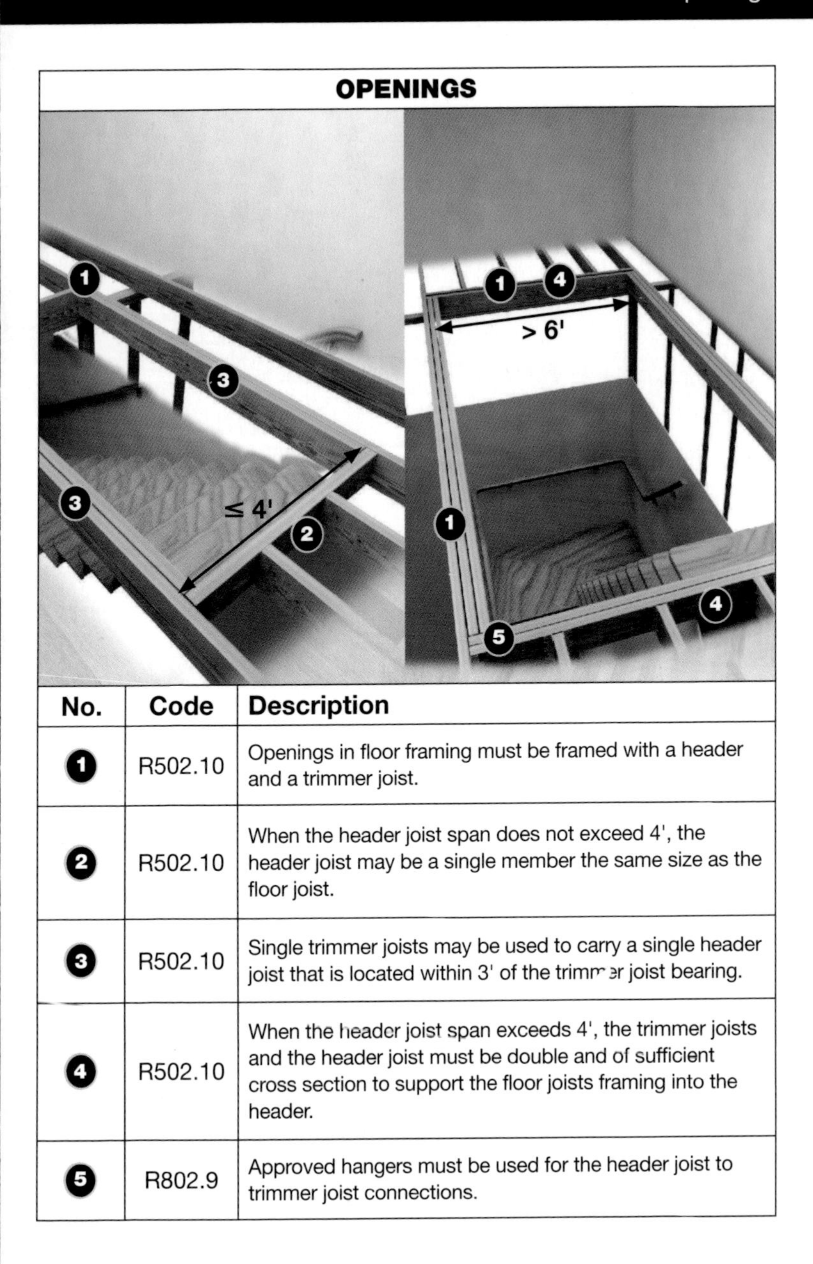

No.	Code	Description
1	R502.10	Openings in floor framing must be framed with a header and a trimmer joist.
2	R502.10	When the header joist span does not exceed 4', the header joist may be a single member the same size as the floor joist.
3	R502.10	Single trimmer joists may be used to carry a single header joist that is located within 3' of the trimmer joist bearing.
4	R502.10	When the header joist span exceeds 4', the trimmer joists and the header joist must be double and of sufficient cross section to support the floor joists framing into the header.
5	R802.9	Approved hangers must be used for the header joist to trimmer joist connections.

CANTILEVER (LIGHT-FRAME BEARING WALL AND ROOF ONLY)

Code	Description
Table R502.3.3(1)(C)	Cantilever backspan must be 3:1
R502.3.3	Span of the cantilever can NOT exceed the nominal depth of the wood floor joist.
Table R502.3.3(1)	The maximum span of the cantilever is based on the table below.

MAXIMUM CANTILEVER SPAN (UPLIFT FORCE AT BACKSPAN SUPPORT IN LBS.)[a]

Member Size and Spacing	Ground Snow Load								
	≤ 20 PSF			30 PSF			50 PSF		
	Roof Width			Roof Width			Roof Width		
	24 ft.	32 ft.	40 ft.	24 ft.	32 ft.	40 ft.	24 ft.	32 ft.	40 ft.
2 × 8 @ 12"	20" (177)	15" (227)	–	18" (209)	–	–	–	–	–
2 × 10 @ 16"	29" (228)	21" (297)	16" (364)	26" (271)	18" (354)	–	20" (375)	–	–
2 × 10 @ 12"	36" (166)	26" (219)	20" (270)	34" (198)	22" (263)	16" (324)	26" (277)	–	–
2 × 12 @ 16"	–	32" (287)	25" (356)	36" (263)	29" (345)	21" (428)	29" (367)	20" (484)	–
2 × 12 @ 12"	–	42" (209)	31" (263)	–	37" (253)	27" (317)	36" (271)	27" (358)	17" (447)
2 × 12 @ 8"	–	48" (136)	45" (169)	–	48" (164)	38" (206)	–	40" (233)	26" (294)

[a] Spans are based on No. 2 Grade lumber of Douglas fir-larch, hem-fir, and spruce-pine-fir for repetitive (three or more) members. No.1 or better shall be used for southern pine.

CANTILEVER (EXTERIOR BALCONY)

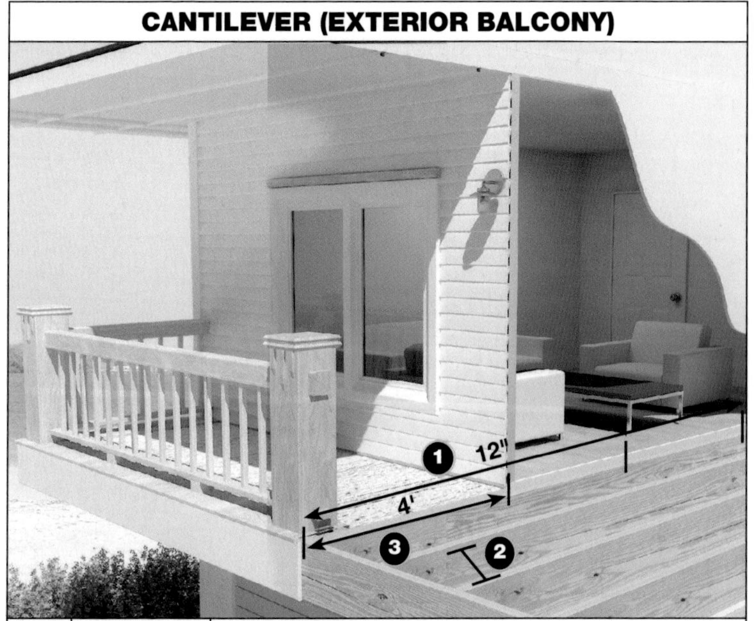

No.	Code	Description
1	R502.3.3(2)	Cantilever backspan must be 2:1. (Table R502.3.3(2) note b)
2	R502.3.3	Span of the cantilever can NOT exceed the nominal depth of the wood floor joist, except as allowed by Tables R502.3.3(1) and/or R502.3.3(2).
3	R502.3.3(2)	The maximum span of the exterior balcony cantilever is based on the table below.

MAXIMUM CANTILEVER SPAN (UPLIFT FORCE AT BACKSPAN SUPPORT IN LBS.)[a]

Member Size and Spacing	Ground Snow Load		
	< 30 PSF	50 PSF	70 PSF
2 × 8 @ 12"	42" (139)	39" (156)	34" (165)
2 × 8 @ 16"	36" (151)	34" (171)	29" (180)
2 × 10 @ 12"	61" (164)	57" (189)	49" (201)
2 × 10 @ 16"	53" (180)	49" (208)	42" (220)
2 × 10 @ 24"	43" (212)	40" (241)	34" (255)
2 × 12 @ 16"	72" (228)	67" (260)	57" (268)
2 × 12 @ 24"	58" (279)	54" (319)	47" (330)

[a] Spans are based on No. 2 Grade lumber of Douglas fir-larch, hem-fir, and spruce-pine-fir for repetitive (three or more) members. No.1 or better shall be used for southern pine.

WALL FRAMING

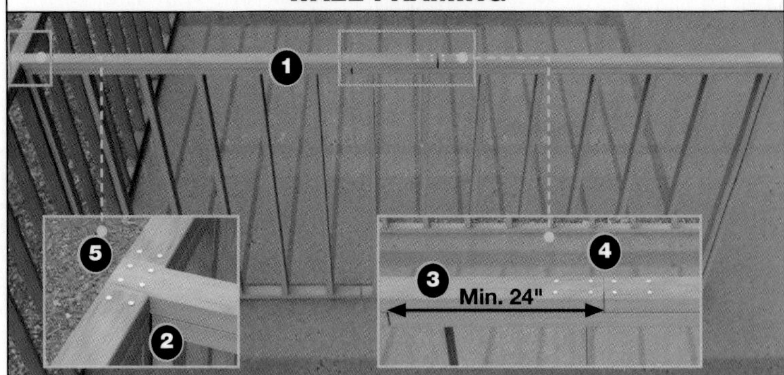

No.	Code	Description
1	R602.3.2	Double top plates are required for wood stud walls.
2	R602.3.2	Plates must provide overlapping at corners and intersections with bearing partitions.
3	R602.3.2	Joints in top plates must be offset at least 24".
	R602.3.2	Joints in top plates are allowed to occur when studs are not directly below.
4	Table R602.3(1)	Lapped area of the required offset for top plates must be nailed with eight 16d common nails. Double plates must be face nailed with 10d nails.
5	Table R602.3(1)	Laps at corners and intersections must be face nailed with three – 10d nails.

SPACING OF WOOD STUDS FOR BEARING WALLS – TABLE R602.3(5)

Lateral Unsupported Height	Member Size	Supporting			
		Roof and Ceiling	1 Floor, Roof, and Ceiling	2 Floors, Roof, and Ceiling	1 Floor
10'	2 × 4	24" o.c	16" o.c	Not allowed	24" o.c
10'	2 × 6	24" o.c	24" o.c	16" o.c	24" o.c

SPACING OF WOOD STUDS FOR NONBEARING WALLS – TABLE R602.3(5)

Lateral Unsupported Height	Member Size	Spacing
14'	2 × 4	24" o.c.
20'	2 × 6	24" o.c.

REQUIRMENTS

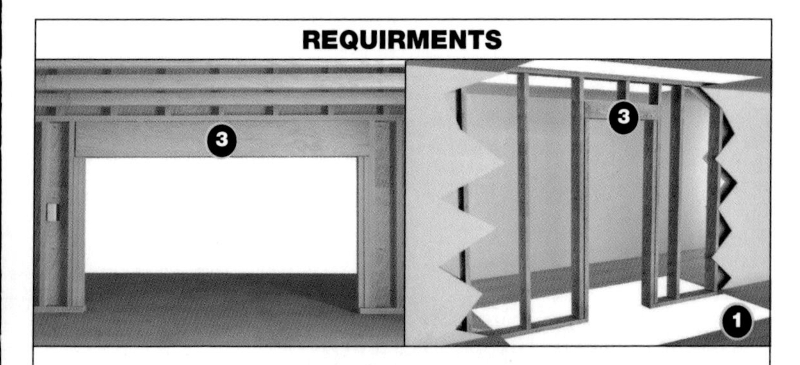

Grade of Veneer on panel back
Grade of Veneer on panel face
Identification Index
Designates the Type of Plywood
Product Standard Governing Manufacturer

STRUCTURAL II
A-C
32 / 16
INTERIOR
PS 1-74
000
EXTERIOR GLUE
APA

Mill Number
Type of Glue Used

No.	Code	Description
1	Tables R503.1, R503.2.1.1(1), and R503.2.1.1(2)	Maximum allowable spans for lumber used as floor sheathing must comply to the following table.
2	R503.2.1	Panel sheathing for structural purposes must be identified by a grade mark of certificate of inspection.
3	Tables R602.7(1) and R602.7(2)	The header span and the number of jacks required to support the header are located on pages 65–68.

MINIMUM THICKNESS OF LUMBER FLOOR SHEATHING		
Joist/Beam Spacing	Perpendicular to Joist	Diagonal to Joist
24"	11/16"	3/4"
16"	5/8"	5/8"

FIREBLOCKING

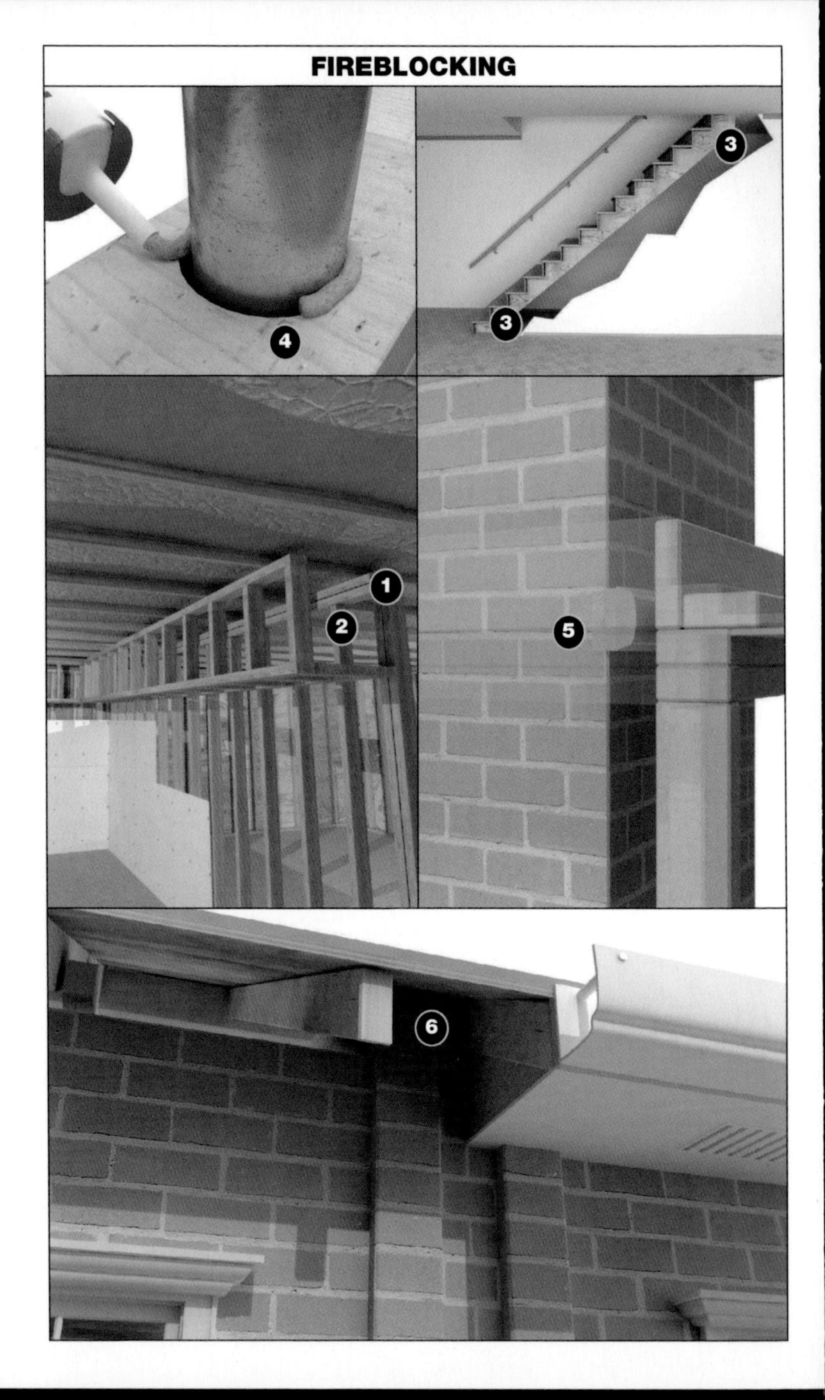

FIREBLOCKING *(cont.)*

No.	Code	Description
1	R302.11	Fireblocking is required vertically at the ceiling and floor levels.
2	R302.11(2)	Fireblocking is required at all interconnections between concealed vertical and horizontal spaces *(soffits, drop ceilings and cove ceilings)*.
3	R302.11(3)	Fireblocking is required in concealed spaces between stair stringers at the top and bottom of the run.
4	R302.11(4)	Fireblocking is required at openings around vents, pipes, ducts, cables, and wires at ceiling and floor level.
5	R302.11(5)	Fireblocking is required for all spaces at floors and ceilings where the chimney passes through.
6	R302.11(6)	Fireblocking is required at the cornice of a two-family dwelling unit at the line of dwelling unit separation.

YOU SHOULD KNOW: R302.11.1

- Fireblocking material includes: 2" nominal lumber, two thicknesses of 1" nominal lumber with broken laps, 23/32" wood structural panels with joints backed by 23/32" wood structural panels, 3/4" particle board with joints backed by 3/4" particle board, 1/2" gypsum board, or 1/4" cement-based millboard.

ROOF FRAMING

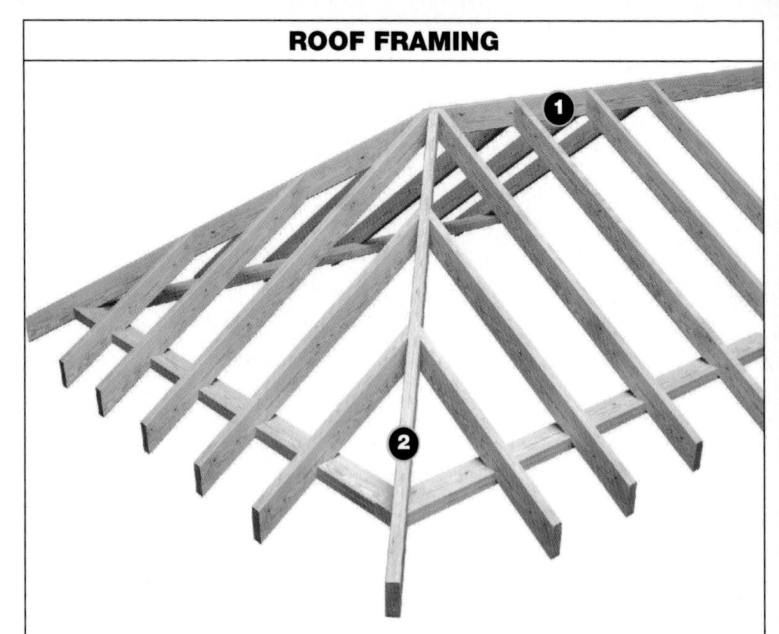

Rafter sizes are based on Table R802.5.1.(1) and (2) on pages 71–75.

No.	Code	Description
1	R802.3	The ridge board must be at least 1" nominal thickness and not less than the cut end of the rafter.
2	R802.3	At all valleys and hips, there must be a valley or hip rafter not less than 2" nominal thickness and not less in depth than the cut end of the rafter.
3	R802.3	Hip and valley rafters must be supported at the ridge by a brace to a bearing partition or be designed to carry and distribute the specific load at that point.
4	R802.3.1	Collar ties must be at least 1" × 4" and attached in the upper third of the attic space.
5	R802.5.1	To reduce the span of rafters, purlins are to be installed and sized no less than the size of the rafters they are to support.
6	R802.5.1	Purlins must be continuous and should be supported by a 2" × 4" brace installed to bearing walls at a slope not less than 45 degrees from the horizontal.

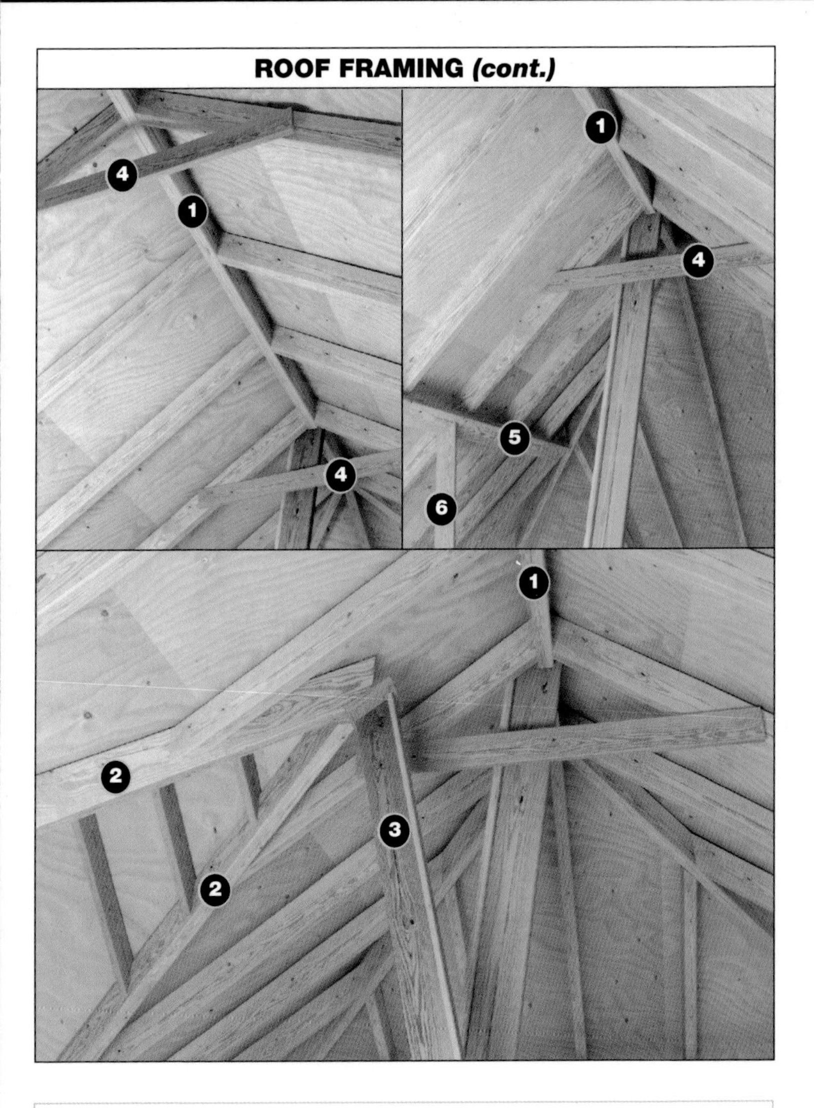

YOU SHOULD KNOW: R802

- Rafters must be framed to ridge board with a gusset plate as a tie.
- Ends of ceiling joists must be lapped a minimum of 3" or butted over bearing partitions or beams and toenailed to the bearing member.
- Ends of rafters or ceiling joist shall have not less than 1 1/2" of bearing on wood or metal *(not less than 3" on masonry or concrete)*.
- Braces for purlins should not be spaced more than 4' on center and the unbraced length of braces can not exceed 8'.

TRUSSES

On-center Spacing Tolerance

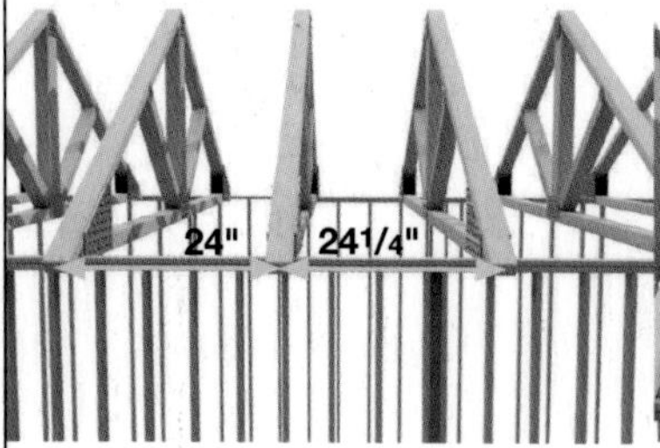

24" 24 1/4"

Out-of-Plumb Tolerance

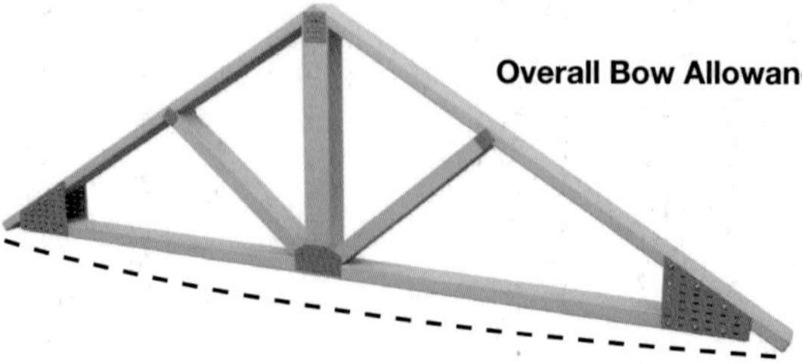

Overall Bow Allowance

Code	Description
R802.10.1	Truss design drawings must be provided to the Building Official and approved prior to installation.
R802.10.3	Trusses must be braced to prevent rotation and provide lateral stability as specified in the drawings.

TECHNICAL STANDARDS BY ANSI/TPI

According to the Truss Institute, trusses must be installed within +/− 1/4" of plan dimensions.

OVERALL BOW ALLOWANCES

Length	150"	175"	200"	225"	250"	275"	300"	350"	400"
Max Bow	3/4"	7/8"	1”	1 1/8"	1 1/4"	1 3/8"	1 1/2"	1 3/4"	2"

MAXIMUM OUT-OF-PLUMB TOLERANCE

Depth	12"	24"	36"	48"	60"	72"	84"	96"	108"
Tolerance	1/4"	1/2"	3/4"	1 "	1 1/4"	1 1/2"	1 3/4"	2"	2"

GYPSUM BOARD

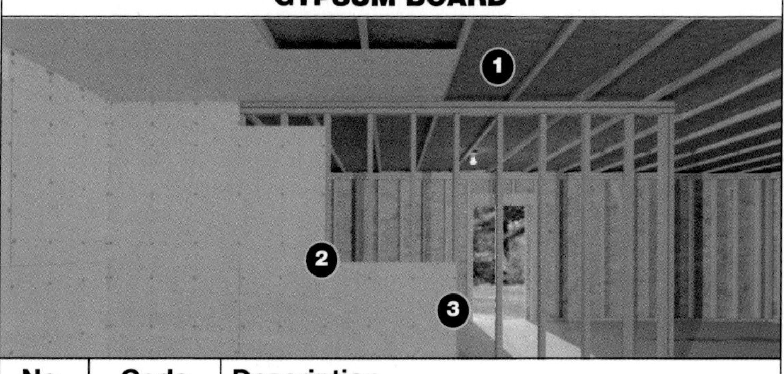

No.	Code	Description
1	R701.2	Install only after all weather tight procedures are in place.
2	R702.3.2	Gypsum board must be supported by 2" nominal material or 1 × 2 furring strips over solid backing.
3	R702.3.5	Gypsum board must be applied at right angles or parallel to framing. All edges and ends must occur on the framing members, except those edges and ends that are perpendicular to the framing members.

1/2" AND 5/8" GYPSUM BOARD FASTENING SCHEDULE - TABLE R702.3.5

Location	Orientation to Framing	Maximum Spacing		
		Framing	Nails	Screws
Without Adhesive				
Ceiling	Perpendicular	24"	7	12
Ceiling	Either Direction	16"	7	12
Walls	Either Direction	24"	8	12
Walls	Either Direction	16"	8	16
With Adhesive				
Ceiling	Either Direction	16"	16	16
Ceiling	Perpendicular	24"	12	16
Walls	Either Direction	24"	16	24

YOU SHOULD KNOW:

- **R109.1.2:** Do not install gypsum board until all rough inspections have been conducted.
- **R702.3.8:** Water Resistant board (Green Board) must NOT be installed over vapor retarder in shower or tub areas. *(For areas requiring green board on ceilings, 12" on center spacing is required on the ceiling joists for 1/2" thickness board.)*

MASONRY VENEER

MASONRY VENEER *(cont.)*		
No.	**Code**	**Description**
1	R703.8.3	Masonry veneer is not to be used to support any load other than the dead load of veneer above it.
2	R703.8.3	Lintels must be installed above openings (see chart for span) and must bear at least 4".
3	R703.8.4	Corrosion-resistant metal ties must be used to anchor veneer to wood backing with nominal 1" separation (up to 4.5" allowed for metal strand wires).
4	R703.8.4.1	Maximum spacing of metal ties is 24" vertically and 32" horizontally. *(Max. 2.67 sq. ft. per tie.)*
5	R703.8.4.1.1	Additional metal ties around the perimeter of openings greater than 16" in either direction, must be spaced not more than 36" on center and placed within 12" of the wall openings.
6	R703.8.5	Flashing must be installed beneath the first course of masonry above finished ground level.
7	R703.8.6	Weepholes (not less than 3/16" diameter) must be provided in the outside wythe of masonry walls, above the flashing, spaced no more than 33" on center.
8	R703.8.2.2	For brick veneer that is to be supported by roof construction, a steel angle placed directly on top of the roof supported by at least three 2" × 6" wood members is required.

SPAN CHART FOR LINTELS SUPPORTING MASONRY VENEER - TABLE R703.8.3.1

Size of Steel Angle	No Story Above	One Story Above	Two Stories Above	Number of 1/2" Reinforcing Bars*
3 × 3 × 1/4	6'	4'6"	3'	1
4 × 3 × 1/4	8'	6'	4'6"	1
5 × 3 1/2 × 5/16	10'	8'	6'	2
6 × 3 1/2 × 5/16	14'	9'6"	7'	2
(2) 6 × 3 1/2 × 5/16	20'	12'	9'6"	4

* applies to reinforced masonry lintels

FLASHING

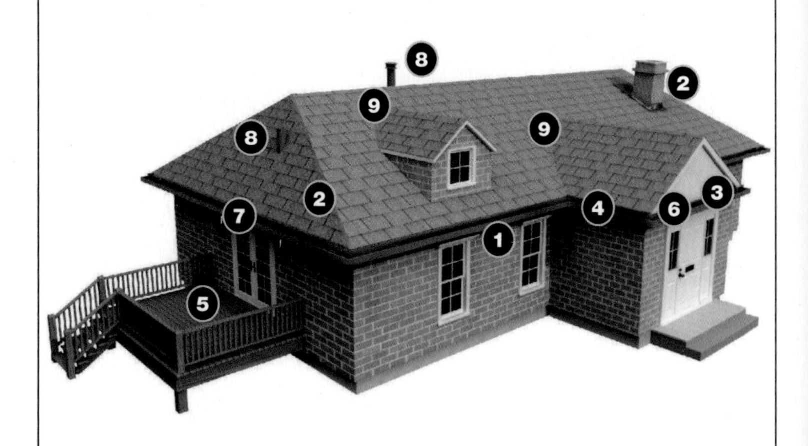

No.	Code	Description
1	R703.4(1)	Flashing must be installed at window and door openings.
2	R703.4(2)	Flashing must be installed at the intersection of chimneys or other masonry construction.
3	R703.4(3)	Flashing must be installed under and at ends of masonry, wood or metal copings and sills.
4	R703.4(4)	Flashing must be installed continuously above all projecting wood trim.
5	R703.4(5)	Flashing must be installed where exterior porches, decks or stairs attach to a wall or floor assembly of wood-frame construction.
6	R703.4(6)	Flashing must be installed at wall and roof intersections
7	R703.4(7)	Flashing must be installed at built-in gutters.
8	R903.2	Flashing must be installed so as to prevent moisture from entering the wall or roof at penetrations through the roof plane.
9	R903.2.1	Flashing must be installed where there is a change in roof slope or change direction.

ROOFING DETAILS

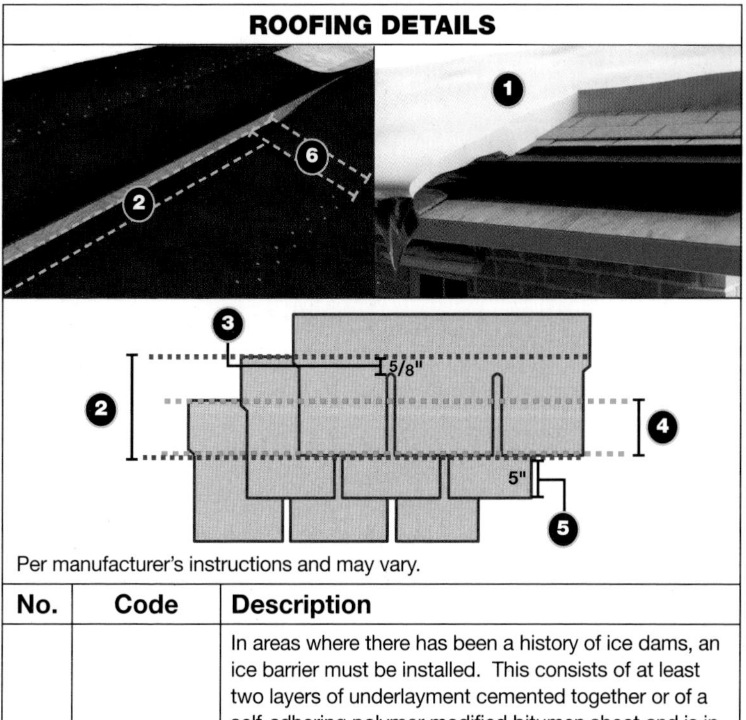

Per manufacturer's instructions and may vary.

No.	Code	Description
1	R905.2.7.1	In areas where there has been a history of ice dams, an ice barrier must be installed. This consists of at least two layers of underlayment cemented together or of a self-adhering polymer modified bitumen sheet and is in lieu of normal underlayment.
		An ice dam is a ridge of ice that forms at the edge of a roof and prevents melting snow (water) from draining off the roof. The water that backs up behind the dam can leak into a home and cause damage to walls, ceilings, insulation, and other areas.
2	—	Top Lap
3	—	Nail Placement
4	—	Head Lap
5	—	Exposure
6	—	End Lap

YOU SHOULD KNOW: R905

- The Code has specific requirements for different roof covering types.

ASPHALT SHINGLES

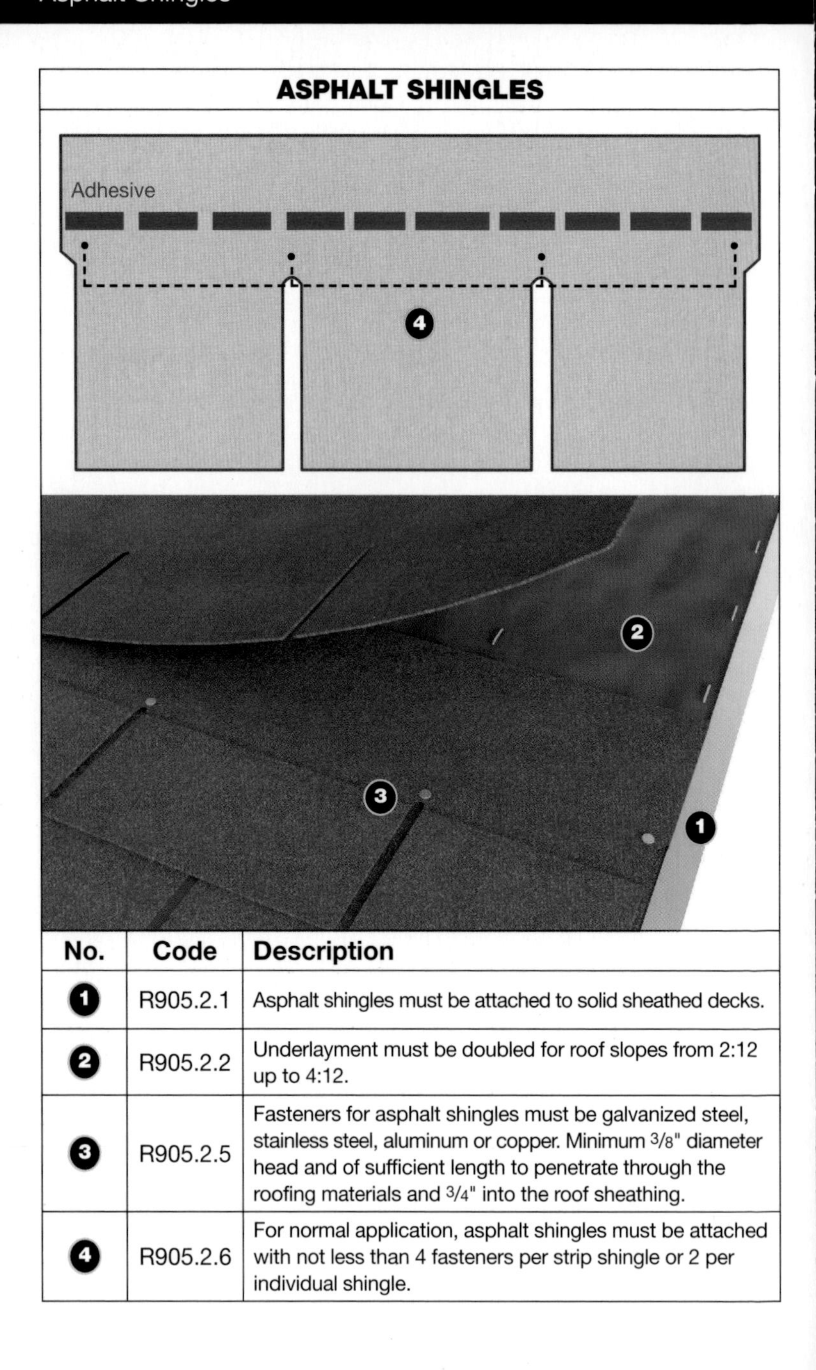

No.	Code	Description
1	R905.2.1	Asphalt shingles must be attached to solid sheathed decks.
2	R905.2.2	Underlayment must be doubled for roof slopes from 2:12 up to 4:12.
3	R905.2.5	Fasteners for asphalt shingles must be galvanized steel, stainless steel, aluminum or copper. Minimum 3/8" diameter head and of sufficient length to penetrate through the roofing materials and 3/4" into the roof sheathing.
4	R905.2.6	For normal application, asphalt shingles must be attached with not less than 4 fasteners per strip shingle or 2 per individual shingle.

CLAY AND CONCRETE TILE

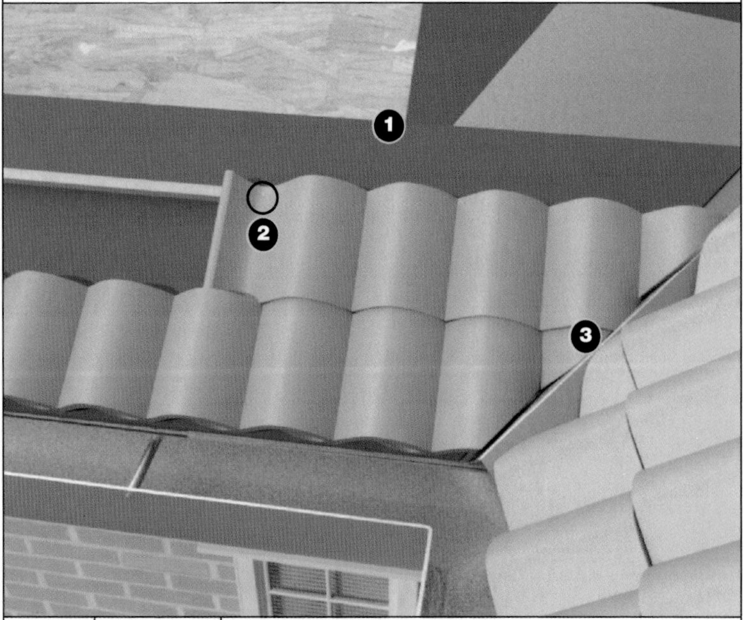

No.	Code	Description
1	R905.3.2	Clay and concrete roof tile are allowed to be installed on roof slopes of $2^1/_2$ units vertical in 12 units horizontal or greater.
	R905.3.2	For roof slopes from $2^1/_2$:12 to 4:12, double underlayment is required.
2	R905.3.6	Nails to fasten clay and concrete roof tiles must be corrosion resistant, 11 gage, $^5/_{16}$" head and must penetrate the deck at least $^3/_4$" or through the thickness of the deck (whichever is less).
3	R905.3.8	Valley flashing must be at least 11" from the center line.

CLAY AND CONCRETE TILE ATTACHMENT - TABLE R905.3.7

Sheathing	Roof Slope	No. of Fasteners
Solid without battens	All	1 per tile
Spaced or solid with battens	< 5:12	Not Required
Spaced sheathing without battens	5:12 < 12:12	1 per tile / every other row
	12:12 < 24:12	1 per tile

WOOD SHINGLES

Code	Description
R905.7.2	Wood shingles are allowed to be installed on slopes of three units vertical in twelve units horizontal or greater.
R905.7.3.1	In areas where there has been a history of ice dams, an ice barrier must be installed extending from the lowest edges of all roof surfaces to a point at least 24" inside the exterior wall line of the building.
R905.7.5	Wood shingles must be laid with a side lap not less than 1 1/2" between joints in courses.
R905.7.5	No two joints in any three adjacent courses of wood shingles are allowed to be in direct alignment.
R905.7.5	Spacing of wood shingles can not be less than 1/4" to 3/8".
R905.7.6	Flashing must be at least #26 gauge (corrosion-resistant sheet metal).
R905.7.6	Flashing must extend 10" from the centerline each way for roofs having slopes less than 12 units vertical in 12 units horizontal and 7" from the centerline each way for slopes of 12 units vertical in 12 units horizontal.
R905.7.6	Flashing must have an end lap of not less than 4".

WOOD SHINGLE WEATHER EXPOSURE AND ROOF SLOPE - TABLE R905.7.5(1)

Shingle Length	Grade	3:12 Slope to < 4:12	4:12 Slope or Steeper
16"	#1	3 3/4"	5"
	#2	3 1/2"	4"
	#3	3"	3 1/2"
18"	#1	4 1/2"	5 1/2"
	#2	4"	4 1/2"
	#3	3 1/2"	4"
24"	#1	5 3/4"	7 1/2"
	#2	5 1/2"	6 1/2"
	#3	5"	5 1/2"

WOOD SHAKES

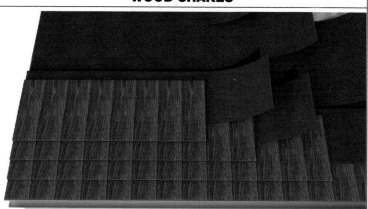

Code	Description
R905.8.2	Wood shakes are allowed to be installed on slopes of three units vertical in twelve units horizontal or greater.
R905.8.3.1	In areas where there has been a history of ice dams, an ice barrier must be installed extending from the lowest edges of all roof surfaces to a point at least 24" inside the exterior wall line of the building.
R905.8.6	Wood shakes must be laid with a side lap not less than 1 1/2" between joints in adjacent courses.
R905.8.6	Spacing between shakes in the same course shall be 3/8" to 5/8" for shakes and tapersawn shakes of naturally durable wood and must be 3/8" to 5/8" preservative treated taper sawn shakes.
R905.8.8	Flashing for wood shake shingles must be at least 26 gauge.
R905.8.8	Flashing for wood shake shingles must extend at least 11" from the centerline each way.
R905.8.8	Flashing must have an end lap of not less than 4".

WOOD SHAKE WEATHER EXPOSURE AND ROOF SLOPE - TABLE R905.8.6

Roofing Material	Grade	Length	4:12 Slope or Steeper
Shakes of Naturally Durable Wood	#1	18"	7 1/2"
	#2	24"	10"
Preservative-Treated Taper-sawn Shakes of Southern Yellow Pine or Taper-sawn shakes of Naturally durable wood	#1	18"	7 1/2"
	#1	24"	10"
	#2	18"	5 1/2"
	#2	24"	7 1/2"

MASONRY CHIMNEYS		
No.	**Code**	**Description**
1	R1003.5	Masonry chimneys can not be corbelled more than 1/2 it's wall thickness (Note: not allowed if the wall or foundation is less than 12" thick unless it projects equally on both sides; however, 2nd story corbelling can equal the wall thickness).
	R1003.5	Projection of a single corbelled course can not exceed 1/2 unit height or 1/3 of the unit bed – whichever is less.
2	R1003.7	The centerline of the flue for offsets constructed with a clay flue liner (one wythe), can NOT extend beyond the center of the chimney below the offset.
3	R1003.20	Crickets to divert water are required for chimneys that are greater than 30" in width.
4	R1003.9	Chimneys must extend at least 2' higher than any portion of a building within 10'.
5	R1003.9	Chimneys can NEVER extend less than 3' above the highest point where they pass through the roof.
6	R1003.9.1	When spark arrestors are installed, they can not be less than 4 times the net free area of the chimney outlet, must have proper heat and corrosion resistance, openings must prohibit passage of spheres 1/2" in diameter but can not block passage of sphere diameter of 3/8".
7	R1009.10	Masonry chimney walls must be constructed of not less than 4" nominal solid masonry units or 4" nominal hollow units grouted solid.
8	R1003.11	Masonry chimneys must be lined with approved flue material types.

YOU SHOULD KNOW: R1003.2

- Footings for masonry chimneys must be at least 12" thick and extend 6" beyond foundation or support wall on all sides.

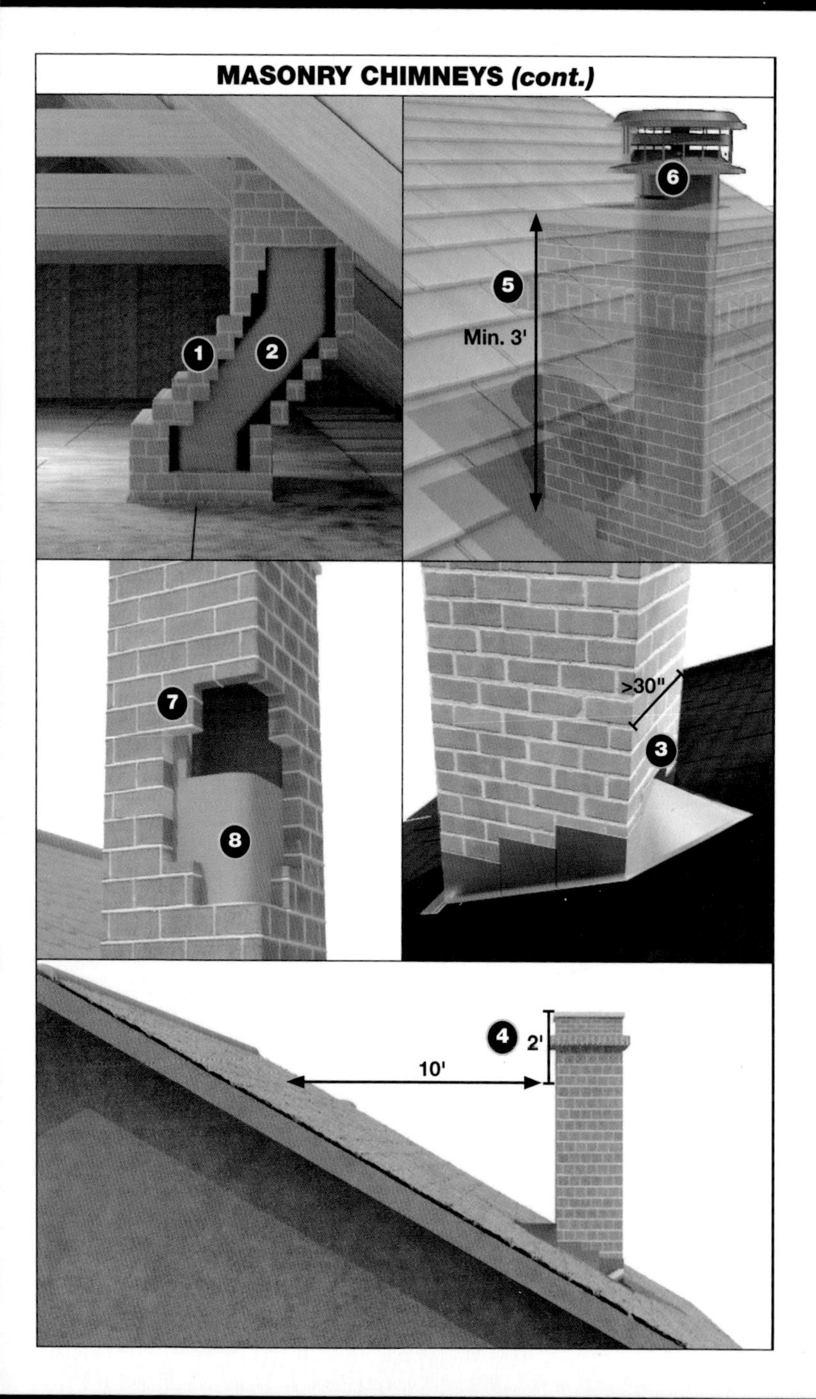
MASONRY CHIMNEYS (cont.)
1
2
6
5
Min. 3'
7
8
>30"
3
4
2'
10'

MASONRY FIREPLACES

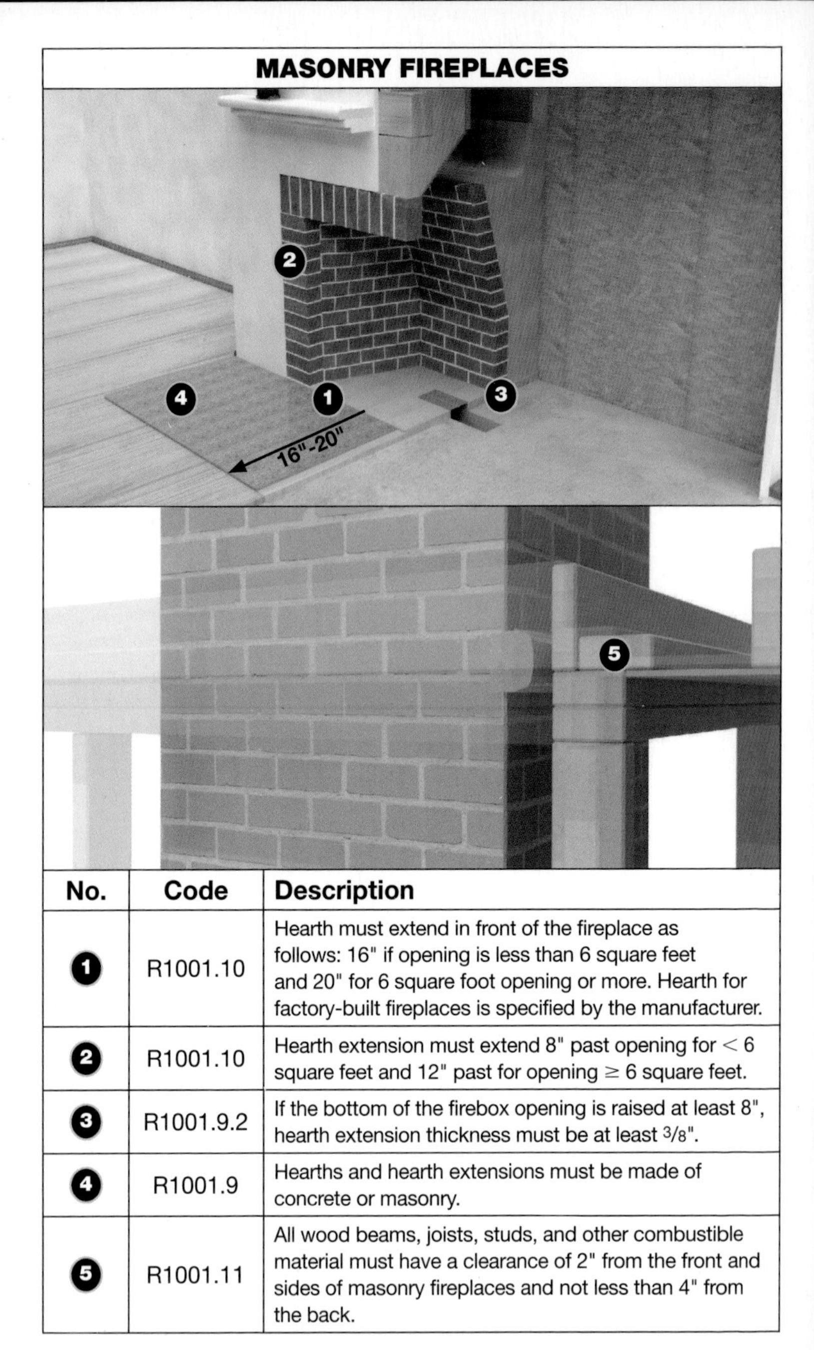

No.	Code	Description
1	R1001.10	Hearth must extend in front of the fireplace as follows: 16" if opening is less than 6 square feet and 20" for 6 square foot opening or more. Hearth for factory-built fireplaces is specified by the manufacturer.
2	R1001.10	Hearth extension must extend 8" past opening for < 6 square feet and 12" past for opening ≥ 6 square feet.
3	R1001.9.2	If the bottom of the firebox opening is raised at least 8", hearth extension thickness must be at least 3/8".
4	R1001.9	Hearths and hearth extensions must be made of concrete or masonry.
5	R1001.11	All wood beams, joists, studs, and other combustible material must have a clearance of 2" from the front and sides of masonry fireplaces and not less than 4" from the back.

ATTIC ACCESS

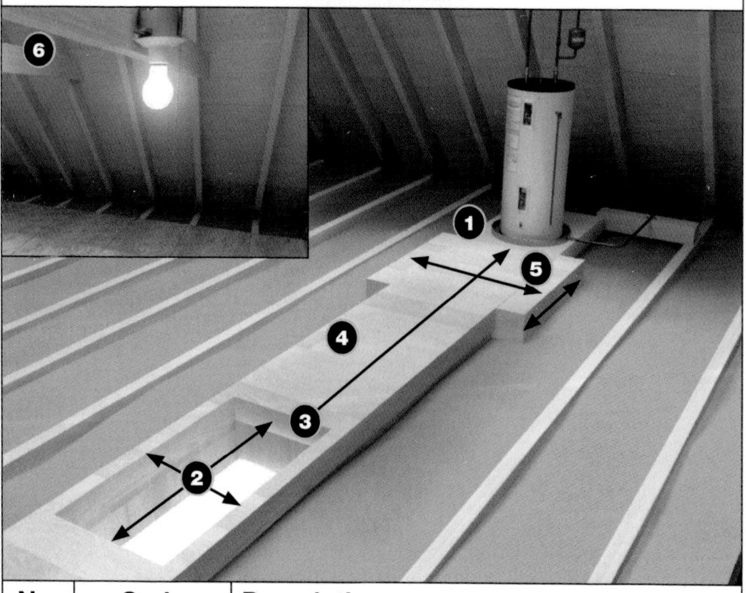

No.	Code	Description
1	M1305.1.3	Appliances must always be installed in locations accessible for maintenance, and replacement.
2	M1305.1.3	Attic access must be at least 22" × 30".
3	M1305.1.3	Access can be no more than 20' from the appliance.
4	M1305.1.3	The path to the appliance must be solid flooring and no less than 24" wide.
5	M1305.1.3	A level work area (service space) at least 30" × 30" must be provided in front of the appliance.
6	M1305.1.3.1	A light controlled by a switch located at the required passageway opening and a receptacle is required.

SKYLIGHTS

No.	Code	Description
1	R308.6	Glazing materials in skylights (including factory assembled), solariums, sunrooms, roofs and sloped walls that are glass, transparent or translucent that are sloped 15 degrees or more from the vertical must comply with the requirements of this section.
2	R308.6.2	Allowable glazing includes: laminated glass, fully tempered glass, heat-strengthened glass, wired glass, and approved rigid plastics.
3	R308.6.8	Unit skylights installed in a roof with a pitch flatter than 3:12 must be mounted on a curb extended at least 4" above the plane of the roof (unless otherwise specified in the manufacturer's installation instructions).

YOU SHOULD KNOW: R308.6.3

- Refer to sections R308.6.3 and R308.6.4 for screen requirements.

PARAPET WALL

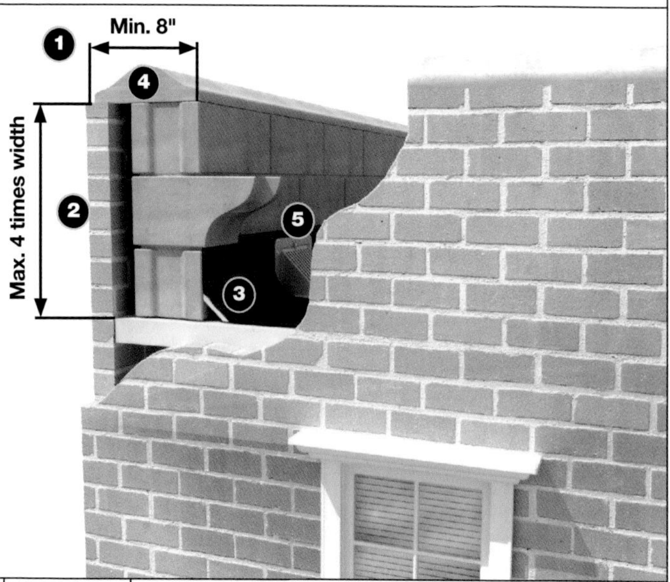

No.	Code	Description
1	R606.2.4	Unreinforced solid masonry parapet walls can be no less than 8" thick (same applies to hollow parapet walls).
2	R606.2.4	Height of unreinforced solid masonry parapet walls can not exceed their width by 4 times (if hollow, height can not exceed width by 3 times).
3	R903.2	Flashing is required at intersections with parapet walls.
4	R903.3	Parapet walls must be properly coped with non-combustible, weatherproof materials of a width not less than the wall thickness.
5	R903.4.1	Overflow drains or scuppers must be installed in parapet walls with the inlet flow located 2" above the low point of the roof served.

MINIMUM VERTICAL REINFORCEMENT FOR 6-INCH NOMINAL FLAT CONCRETE BASEMENT WALLS

Max. Wall Ht.	Max. Unbalanced Backfill Ht.	GW, GP, SW and SP Soils 30		GM, GC, SM-SC and ML Soils 45		SC, MH, ML-CL and CL Soils 60	
		Min. Size	Reinforcement Bar Spacing	Min. Size	Reinforcement Bar Spacing	Min. Size	Reinforcement Bar Spacing
8'	4'						
	5'			6	39"	6	48"
	6'	5	39"	6	48"	6	35"
	7'	6	48"	6	34"	6	25"
	8'	6	39"	6	25"	6	18"
9'	4'						
	5'			5	37"	6	48"
	6'	5	36"	6	44"	6	32"
	7'	6	47"	6	30"	6	22"
	8'	6	34"	6	22"	6	16"
	9'	6	27"	6	17"		
10'	4'	NR	NR	NR	NR	NR	NR
	5'	NR	NR	5	35"	6	48"
	6'	6	48"	6	41"	6	30"
	7'	6	43"	6	28"	6	20"
	8'	6	31"	6	20"	DR	DR
	9'	6	24"	6	15"	DR	DR
	10'	6	19"	DR	DR	DR	DR

FLOOR JOIST SPAN CHART: RESIDENTIAL SLEEPING AREA LIVE LOAD = 30 PSF, DEAD LOAD = 10 PSF (L/Δ=360) BASED ON 16" JOIST SPACING					
Species	**Grade**	**2 × 6**	**2 × 8**	**2 × 10**	**2 × 12**
Southern Pine	#1	10'9"	14'2"	18'0"	21'4"
	#2	10'3"	13'3"	15'8"	18'6"
	#3	7'11"	10'0"	11'1"	14'4"
Spruce/Pine/Fir	#1	10'3"	13'6"	17'2"	19'11"
	#2	10'3"	13'6"	17'2"	19'11"
	#3	8'8"	11'0"	13'5"	15'7"
Douglas Fir-Larch	#1	10'11"	14'5"	18'5"	21'4"
	#2	10'9"	14'2"	17'5"	20'3"
	#3	8'7"	10'11"	13'4"	15'5"
Hem-Fir	#1	10'6"	13'10"	17'8"	21'1"
	#2	10'0"	13'2"	16'10"	19'8"
	#3	8'5"	10'8"	13'0"	15'1"

FLOOR JOIST SPAN CHART: RESIDENTIAL SLEEPING AREA LIVE LOAD = 30 PSF, DEAD LOAD = 20 PSF (L/Δ=360) BASED ON 16" JOIST SPACING					
Species	**Grade**	**2 × 6**	**2 × 8**	**2 × 10**	**2 × 12**
Southern Pine	#1	10'9"	13'9"	16'1"	19'1"
	#2	9'4"	11'10"	14'0"	16'6"
	#3	7'1"	8'11"	10'10"	12'10"
Spruce/Pine/Fir	#1	9'11"	12'7"	15'5"	17'10"
	#2	9'11"	12'7"	15'5"	17'10"
	#3	7'6"	9'6"	11'8"	13'6"
Douglas Fir-Larch	#1	10'8"	13'6"	16'5"	19'1"
	#2	10'1"	12'9"	15'7"	18'1"
	#3	7'8"	9'9"	11'11"	13'10"
Hem-Fir	#1	10'6"	13'4"	16'3"	18'10"
	#2	9'10"	12'5"	15'2"	17'7"
	#3	7'6"	9'6"	11'8"	13'6"

FLOOR JOIST SPAN CHART: RESIDENTIAL SLEEPING AREA LIVE LOAD = 40 PSF, DEAD LOAD = 10 PSF (L/Δ=360) BASED ON 16" JOIST SPACING					
Species	**Grade**	**2 × 6**	**2 × 8**	**2 × 10**	**2 × 12**
Southern Pine	#1	9'9"	12'10"	16'1"	19'1"
	#2	9'4"	11'10"	14'0"	16'6"
	#3	7'1"	8'11"	10'10"	12'10"
Spruce/Pine/Fir	#1	9'4"	12'3"	15'5"	17'10"
	#2	9'4"	12'3"	15'5"	17'10"
	#3	7'6"	9'6"	11'8"	13'6"
Douglas Fir-Larch	#1	9'11"	13'1"	16'5"	19'1"
	#2	9'9"	12'9"	15'7"	18'1"
	#3	7'8"	9'9"	11'11"	13'10"
Hem-Fir	#1	9'6"	12'7"	16'0"	18'10"
	#2	9'1"	12'0"	15'2"	17'7"
	#3	7'6"	9'6"	11'8"	13'6"

FLOOR JOIST SPAN CHART: RESIDENTIAL SLEEPING AREA LIVE LOAD = 40 PSF, DEAD LOAD = 20 PSF (L/Δ=360) BASED ON 16" JOIST SPACING					
Species	**Grade**	**2 × 6**	**2 × 8**	**2 × 10**	**2 × 12**
Southern Pine	#1	10'9"	13'9"	16'1"	19'1"
	#2	9'4"	11'10"	14'0"	16'6"
	#3	7'1"	8'11"	10'10"	12'10"
Spruce/Pine/Fir	#1	9'1"	11'6"	14'1"	16'3"
	#2	9'1"	11'6"	14'1"	16'3"
	#3	6'10"	8'8"	10'7"	12'4"
Douglas Fir-Larch	#1	9'8"	12'4"	15'0"	17'5"
	#2	9'8"	12'4"	15'0"	17'5"
	#3	9'3"	11'8"	14'3"	16'6"
Hem-Fir	#1	7'0"	8'11"	10'11"	12'7"
	#2	8'11"	11'4"	13'10"	16'1"
	#3	6'10"	8'8"	10'7"	12'4"

GIRDER/HEADER SPANS FOR EXTERIOR BEARING WALLS (30PSF GROUND SNOW LOAD) (#2 DOUGLAS FIR-LARCH, HEM-FIR, SPRUCE-PINE-FIR, AND NO. 1 SOUTHERN PINE)

		Building Width					
		20'		28'		36'	
Supporting	Size	Span	Jacks	Span	Jacks	Span	Jacks
Roof & Ceiling	2–2 × 4	3'6"	1	3'2"	1	2'10"	1
	2–2 × 6	5'5"	1	4'8"	1	4'2"	1
	2–2 × 8	6'10"	1	5'11"	2	5'4"	2
	2–2 × 10	8'5"	2	7'3"	2	6'6"	2
	2–2 × 12	9'9"	2	8'5"	2	7'6"	2
	3–2 × 8	8'4"	1	7'5"	1	6'8"	1
	3–2 × 10	10'6"	1	9'1"	2	8'2"	2
	3–2 × 12	12'2"	2	10'7"	2	9'5"	2
Roof, Ceiling & One Center-Bearing Floor	2–2 × 4	3'1"	1	2'9"	1	2'5"	1
	2–2 × 6	4'6"	1	4'0"	1	3'7"	2
	2–2 × 8	5'9"	2	5'0"	2	4'6"	2
	2–2 × 10	7'0"	2	6'2"	2	5'6"	2
	2–2 × 12	8'1"	2	7'1"	2	6'5"	2
	3–2 × 8	7'2"	1	6'3"	2	5'8"	2
	3–2 × 10	8'9"	2	7'8"	2	6'11"	2
	3–2 × 12	10'2"	2	8'11"	2	8'0"	2

2015 IRC®, International Code Council®

GIRDER/HEADER SPANS FOR EXTERIOR BEARING WALLS (30PSF GROUND SNOW LOAD) (#2 DOUGLAS FIR-LARCH, HEM-FIR, SPRUCE-PINE-FIR, AND NO. 1 SOUTHERN PINE) *(cont.)*

Supporting	Size	Building Width					
		20'		28'		36'	
		Span	Jacks	Span	Jacks	Span	Jacks
Roof, Ceiling & One Clear Span Floor	2–2 × 4	2'8"	1	2'4"	1	2'1"	1
	2–2 × 6	3'11"	1	3'5"	2	3'0"	2
	2–2 × 8	5'0"	2	4'4"	2	3'10"	2
	2–2 × 10	6'1"	2	5'3"	2	4'8"	2
	2–2 × 12	7'1"	2	6'1"	3	5'5"	3
	3–2 × 8	6'3"	2	5'5"	2	4'10"	2
	3–2 × 10	7'7"	2	6'7"	2	5'11"	2
	3–2 × 12	8'10"	2	7'8"	2	6'10"	2
Roof, Ceiling & Two Center-Bearing Floors	2–2 × 4	2'7"	1	2'3"	1	2'0"	1
	2–2 × 6	3'9"	2	3'3"	2	2'11"	2
	2–2 × 8	4'9"	2	4'2"	2	3'9"	2
	2–2 × 10	5'9"	2	5'1"	2	4'7"	3
	2–2 × 12	6'8"	2	5'10"	3	5'3"	3
	3–2 × 8	5'11"	2	5'2"	2	4'8"	2
	3–2 × 10	7'3"	2	6'4"	2	5'8"	2
	3–2 × 12	8'5"	2	7'4"	2	6'7"	2

GIRDER/HEADER SPANS FOR EXTERIOR BEARING WALLS (30PSF GROUND SNOW LOAD) (#2 DOUGLAS FIR-LARCH, HEM-FIR, SPRUCE-PINE-FIR, AND NO. 1 SOUTHERN PINE) *(cont.)*

		Building Width					
		20'		28'		36'	
Supporting	Size	Span	Jacks	Span	Jacks	Span	Jacks
Roof, Ceiling & Two Clear Span Floor	2–2 × 4	2'1"	1	1'8"	1	1'6"	2
	2–2 × 6	3'1"	2	2'8"	2	2'4"	2
	2–2 × 8	3'10"	2	3'4"	2	3'0"	3
	2–2 × 10	4'9"	2	4'1"	3	3'8"	3
	2–2 × 12	5'6"	3	4'9"	3	4'3"	3
	3–2 × 8	4'10"	2	4'2"	2	3'9"	2
	3–2 × 10	5'11"	2	5'1"	2	4'7"	3
	3–2 × 12	6'10"	2	5'11"	3	5'4"	3

2015 IRC®, International Code Council®

GIRDER/HEADER SPANS FOR INTERIOR BEARING WALLS (#2 DOUGLAS FIR-LARCH, HEM-FIR, SPRUCE-PINE-FIR, AND NO. 1 SOUTHERN PINE) *(cont.)*

		Building Width					
		20'		28'		36'	
Supporting	Size	Span	Jacks	Span	Jacks	Span	Jacks
1 floor	2–2 × 4	3'1"	1	2'8"	1	2'5"	1
	2–2 × 6	4'6"	1	3'11"	1	3'6"	1
	2–2 × 8	5'9"	1	5'0"	2	4'5"	2
	2–2 × 10	7'0"	2	6'1"	2	5'5"	2
	2–2 × 12	8'1"	2	7'0"	2	6'3"	2
	3–2 × 8	7'2"	1	6'3"	1	5'7"	2
	3–2 × 10	8'9"	1	7'7"	2	6'9"	2
	3–2 × 12	10'2"	2	8'10"	2	7'10"	2
2 floors	2–2 × 4	2'2"	1	1'10"	1	1'7"	1
	2–2 × 6	3'2"	2	2'9"	2	2'5"	2
	2–2 × 8	4'1"	2	3'6"	2	3'2"	2
	2–2 × 10	4'11"	2	4'3"	2	3'10"	3
	2–2 × 12	5'9"	2	5'0"	3	4'5"	3
	3–2 × 8	5'1"	2	4'5"	2	3'11"	2
	3–2 × 10	6'2"	2	5'4"	2	4'10"	2
	3–2 × 12	7'2"	2	6'3"	2	5'7"	3

CEILING JOIST SPAN CHART (FOR ATTICS WITHOUT STORAGE — LIVE LOAD = 10 PSF, DEAD LOAD = 5 PSF)

		Member Size and Spacing							
		2 × 4		2 × 6		2 × 8		2 × 10	
Species	Grade	16"	24"	16"	24"	16"	24"	16"	24"
Southern Pine	#1	11'3"	9'10"	17'8"	15'6"	23'10"	20'5"	—	24'0"
	#2	10'9"	9'3"	16'11"	13'11"	21'7"	17'7"	25'7"	20'11"
	#3	8'9"	7'2"	12'11"	10'6"	16'3"	13'3"	19'9"	16'1"
Spruce/Pine/Fir	#1	10'9"	9'5"	16'11"	14'9"	22'4"	18'9"	—	22'11"
	#2	10'9"	9'5"	16'11"	14'9"	22'4"	18'9"	—	22'11"
	#3	9'5"	7'8"	13'9"	11'2"	17'5"	14'2"	21'3"	17'4"
Douglas Fir-Larch	#1	11'6"	10'6"	18'1"	15'9"	23'10"	20'1"	—	24'6"
	#2	11'3"	9'10"	17'6"	15'0"	23'4"	19'1"	—	23'3"
	#3	9'7"	7'10"	14'1"	11'6"	17'10"	14'7"	21'9"	17'9"
Hem-Fir	#1	11'0"	9'8"	17'4"	15'2"	22'10"	19'10"	—	24'3"
	#2	10'6"	9'2"	16'6"	14'5"	21'9"	18'6"	—	22'7"
	#3	9'5"	7'8"	13'9"	11'2"	17'5"	14'2"	21'3"	17'4"

CEILING JOIST SPAN CHART (FOR ATTICS WITH LIMITED STORAGE — LIVE LOAD = 20 PSF, DEAD LOAD = 10 PSF)									
		Member Size and Spacing							
		2 × 4		2 × 6		2 × 8		2 × 10	
Species	**Grade**	**16"**	**24"**	**16"**	**24"**	**16"**	**24"**	**16"**	**24"**
Southern Pine	#1	8'11"	7'8"	14'0"	11'5"	17'9"	14'6"	20'9"	16'11"
	#2	8'0"	6'7"	12'0"	9'10"	17'9"	12'6"	20'9"	14'9"
	#3	6'2"	5'8"	9'2"	8'4"	11'6"	10'6"	14'0"	12'9"
Spruce/Pine/Fir	#1	8'7"	7'2"	12'10"	10'6"	16'3"	13'3"	19'10"	16'3"
	#2	8'7"	7'2"	12'10"	10'6"	16'3"	13'3"	19'10"	16'3"
	#3	6'8"	5'5"	9'8"	7'11"	12'4"	10'0"	15'0"	12'3"
Douglas Fir-Larch	#1	9'1"	7'8"	13'9"	11'2"	17'5"	14'2"	21'3"	17'4"
	#2	8'11"	7'3"	13'0"	10'8"	15'6"	13'6"	20'2"	16'5"
	#3	8'11"	5'7"	13'0"	8'1"	15'6"	10'3"	20'2"	12'7"
Hem-Fir	#1	8'9"	7'7"	13'7"	11'1"	17'2"	14'0"	14'0"	17'1"
	#2	8'4"	7'1"	12'8"	10'4"	16'0"	13'1"	19'7"	16'0"
	#3	6'8"	5'5"	12'8"	7'11"	12'4"	10'0"	15'0"	12'3"

RAFTER SPAN CHART (CEILING NOT ATTACHED TO RAFTERS L / Δ = 180) ROOF LIVE LOAD = 20 PSF, DEAD LOAD = 10 PSF											
		Member Size and Spacing									
		2 × 4		2 × 6		2 × 8		2 × 10		2 × 12	
Species	**Grade**	**16"**	**24"**	**16"**	**24"**	**16"**	**24"**	**16"**	**24"**	**16"**	**24"**
Southern Pine	#1	9'10"	8'7"	15'6"	12'9"	19'10"	16'2"	23'10"	18'11"	—	22'6"
	#2	9'0"	7'4"	13'6"	11'0"	17'1"	10'11"	20'3"	16'6"	23'10"	19'6"
	#3	6'11"	5'8"	10'2"	8'4"	12'10"	10'6"	15'7"	12'9"	18'6"	15'1"
Spruce/ Pine/Fir	#1	9'5"	8'0"	14'4"	11'9"	18'2"	14'10"	22'3"	18'2"	25'9"	21'0"
	#2	9'5"	8'0"	14'4"	11'9"	18'2"	14'10"	22'3"	18'2"	25'9"	21'0"
	#3	7'5"	6'1"	10'10"	8'10"	13'9"	11'3"	16'9"	13'8"	19'6"	15'11"
Douglas Fir-Larch	#1	9'5"	8'7"	14'0"	12'6"	17'9"	15'10"	21'8"	19'5"	18'6"	22'6"
	#2	9'1"	8'2"	13'3"	11'11"	16'10"	15'1"	20'7"	18'5"	23'10"	21'4"
	#3	6'11"	6'2"	10'2"	9'11"	12'10"	11'6"	15'8"	14'11"	18'3"	16'3"
Hem-Fir	#1	9'8"	8'5"	15'2"	12'4"	19'3"	15'8"	23'5"	19'2"	—	22'2"
	#2	9'2"	7'11"	14'2"	11'7"	17'11"	14'8"	21'11"	17'10"	25'5"	20'9"
	#3	7'5"	6'1"	10'10"	8'10"	13'9"	11'3"	16'9"	13'8"	19'6"	15'11"

2015 IRC®, International Code Council®

RAFTER SPAN CHART (CEILING NOT ATTACHED TO RAFTERS L / Δ = 180) ROOF LIVE LOAD = 20 PSF, DEAD LOAD = 20 PSF											
		Member Size and Spacing									
		2 × 4		2 × 6		2 × 8		2 × 10		2 × 12	
Species	**Grade**	**16"**	**24"**	**16"**	**24"**	**16"**	**24"**	**16"**	**24"**	**16"**	**24"**
Southern Pine	#1	9'1"	7'5"	13'7"	11'1"	17'2"	14'0"	20'1"	16'5"	23'10"	19'6"
	#2	7'9"	6'4"	11'8"	9'6"	14'9"	12'1"	17'6"	14'4"	20'8"	16'10"
	#3	6'0"	4'11"	8'10"	7'3"	11'2"	9'1"	13'6"	11'0"	16'0"	13'1"
Spruce/ Pine/Fir	#1	8'6"	6'11"	12'5"	10'2"	15'9"	12'10"	19'3"	15'8"	22'4"	18'3"
	#2	8'6"	6'11"	12'5"	10'2"	15'9"	12'10"	19'3"	15'8"	22'4"	18'3"
	#3	6'5"	5'3"	9'5"	7'8"	11'11"	9'9"	14'6"	11'10"	16'10"	13'9"
Douglas Fir-Larch	#1	8'4"	7'4"	12'2"	10'9"	15'4"	13'7"	18'9"	16'7"	21'9"	19'3"
	#2	7'10"	6'10"	11'6"	10'0"	14'7"	12'8"	17'10"	15'6"	20'8"	17'11"
	#3	6'7"	5'3"	8'9"	7'8"	11'2"	9'9"	12'7"	11'10"	15'0"	13'9"
Hem-Fir	#1	8'2"	7'4"	12'0"	10'9"	15'2"	13'7"	18'6"	16'7"	21'6"	19'3"
	#2	7'8"	6'10"	11'2"	10'0"	14'2"	12'8"	17'4"	15'6"	20'1"	17'11"
	#3	5'10"	5'3"	8'7"	7'8"	10'10"	9'9"	13'3"	11'10"	15'5"	13'9"

RAFTER SPAN CHART (CEILING ATTACHED TO RAFTERS L / Δ = 240)
ROOF LIVE LOAD = 20 PSF, DEAD LOAD = 10 PSF

		Member Size and Spacing									
		2 × 4		2 × 6		2 × 8		2 × 10		2 × 12	
Species	**Grade**	**16"**	**24"**	**16"**	**24"**	**16"**	**24"**	**16"**	**24"**	**16"**	**24"**
Southern Pine	#1	8'11"	7'10"	16'1"	12'3"	18'6"	16'2"	23'2"	18'11"	—	22'6"
	#2	8'7"	7'4"	13'5"	11'0"	17'1"	13'11"	20'3"	16'6"	23'10"	19'6"
	#3	6'11"	5'8"	10'2"	8'4"	12'10"	10'6"	15'7"	12'9"	18'6"	15'1"
Spruce/Pine/Fir	#1	8'7"	7'6"	13'5"	11'9"	17'9"	14'10"	22'3"	18'2"	25'9"	21'0"
	#2	8'7"	7'6"	13'5"	11'9"	17'9"	14'10"	22'3"	18'2"	25'9"	21'0"
	#3	7'5"	6'1"	10'10"	8'10"	13'9"	11'3"	16'9"	13'8"	19'6"	15'11"
Douglas Fir-Larch	#1	9'1"	8'0"	14'4"	12'6"	18'11"	15'10"	23'9"	19'5"	—	22'6"
	#2	8'11"	7'10"	14'1"	11'9"	18'5"	14'10"	22'6"	18'2"	26'0"	21'0"
	#3	7'7"	6'2"	11'1"	9'1"	14'1"	11'6"	17'2"	14'1"	19'11"	16'3"
Hem-Fir	#1	8'9"	7'8"	13'9"	12'0"	18'1"	15'8"	23'1"	19'2"	—	22'2"
	#2	8'4"	7'3"	13'1"	11'5"	17'3"	14'8"	21'11"	17'10"	25'5"	20'9"
	#3	7'5"	6'1"	10'10"	8'10"	13'9"	11'3"	16'9"	13'8"	19'6"	15'11"

2015 IRC®, International Code Council®

RAFTER SPAN CHART (CEILING ATTACHED TO RAFTERS L / Δ = 240)
ROOF LIVE LOAD = 20 PSF, DEAD LOAD = 20 PSF

		Member Size and Spacing									
		2 × 4		2 × 6		2 × 8		2 × 10		2 × 12	
Species	Grade	16"	24"	16"	24"	16"	24"	16"	24"	16"	24"
Southern Pine	#1	8'11"	8'1"	13'7"	12'9"	17'2"	16'10"	20'1"	20'10"	23'10"	24'8"
	#2	7'9"	7'5"	11'8"	11'1"	14'9"	14'0"	17'6"	16'5"	20'8"	19'6"
	#3	6'0"	4'11"	8'10"	7'3"	11'2"	9'1"	13'6"	11'0"	16'0"	13'1"
Spruce/ Pine/Fir	#1	8'6"	6'11"	12'5"	10'2"	15'9"	12'10"	19'3"	15'8"	22'4"	18'3"
	#2	8'6"	6'11"	12'5"	10'2"	15'9"	12'10"	19'3"	15'8"	22'4"	18'3"
	#3	6'5"	5'3"	9'5"	7'8"	11'11"	9'9"	14'6"	11'10"	16'10"	13'9"
Douglas Fir-Larch	#1	9'1"	7'5"	13'3"	10'10"	16'10"	13'9"	20'7"	16'9"	23'10"	19'6"
	#2	8'7"	7'0"	12'7"	10'4"	16'0"	13'0"	19'6"	15'11"	22'7"	18'6"
	#3	6'7"	5'4"	9'8"	7'10"	12'2"	10'0"	14'11"	12'2"	17'3"	14'1"
Hem-Fir	#1	8'9"	7'4"	13'1"	10'9"	16'7"	13'7"	20'4"	16'7"	23'7"	19'3"
	#2	8'4"	6'10"	12'3"	10'0"	15'6"	12'8"	18'11"	15'6"	22'0"	17'11"
	#3	6'5"	5'3"	9'5"	7'8"	11'11"	9'9"	14'6"	11'10"	16'10"	13'9"

2015 IRC®, International Code Council®

RAFTER SPAN CHART (CEILING NOT ATTACHED TO RAFTERS L / Δ = 180)
GROUND SNOW LOAD = 30 PSF, DEAD LOAD = 10 PSF

		Member Size and Spacing									
		2 × 4		2 × 6		2 × 8		2 × 10		2 × 12	
Species	**Grade**	**16"**	**24"**	**16"**	**24"**	**16"**	**24"**	**16"**	**24"**	**16"**	**24"**
Southern Pine	#1	8'7"	7'1"	13'0"	10'7"	16'6"	13'5"	19'3"	15'9"	22'10"	18'8"
	#2	7'6"	6'1"	11'2"	9'2"	14'2"	11'7"	16'10"	13'9"	19'10"	16'2"
	#3	5'9"	4'8"	8'6"	6'11"	10'8"	8'9"	13'0"	10'7"	15'4"	12'6"
Spruce/ Pine/Fir	#1	8'2"	6'8"	11'11"	9'9"	15'1"	12'4"	18'5"	15'1"	21'5"	17'6"
	#2	8'2"	6'8"	11'11"	9'9"	15'1"	12'4"	18'5"	15'1"	21'5"	17'6"
	#3	6'2"	5'0"	9'0"	7'4"	11'5"	9'4"	13'11"	11'5"	16'2"	13'2"
Douglas Fir-Larch	#1	8'9"	7'1"	12'9"	10'5"	16'2"	13'2"	19'9"	16'1"	22'10"	18'8"
	#2	8'3"	6'9"	12'1"	9'10"	15'4"	12'6"	18'9"	15'3"	21'8"	17'9"
	#3	6'4"	5'2"	9'3"	7'7"	11'8"	9'7"	14'3"	11'8"	16'7"	13'6"
Hem-Fir	#1	8'5"	7'0"	12'7"	10'3"	15'11"	13'3"	19'6"	15'11"	22'7"	18'5"
	#2	8'0"	6'7"	11'9"	9'7"	14'11"	12'2"	18'2"	14'10"	21'1"	17'3"
	#3	6'2"	5'0"	9'0"	7'4"	11'5"	9'4"	13'11"	11'5"	16'2"	13'2"

RAFTER SPAN CHART (CEILING NOT ATTACHED TO RAFTERS L / Δ = 180) GROUND SNOW LOAD = 30 PSF, DEAD LOAD = 20 PSF											
		Member Size and Spacing									
		2 × 4		2 × 6		2 × 8		2 × 10		2 × 12	
Species	**Grade**	**16"**	**24"**	**16"**	**24"**	**16"**	**24"**	**16"**	**24"**	**16"**	**24"**
Southern Pine	#1	7'10"	6'4"	11'7"	9'6"	14'9"	12'0"	17'3"	14'1"	20'5"	16'8"
	#2	6'8"	5'5"	10'0"	8'2"	12'8"	10'4"	15'1"	12'3"	17'9"	14'6"
	#3	5'2"	4'2"	7'7"	6'2"	9'7"	7'10"	11'7"	9'6"	13'9"	11'2"
Spruce/ Pine/Fir	#1	7'3"	5'11"	10'8"	8'8"	13'6"	11'0"	16'6"	13'6"	19'2"	15'7"
	#2	7'3"	5'11"	10'8"	8'8"	13'6"	11'0"	16'6"	13'6"	19'2"	15'7"
	#3	5'6"	4'6"	8'1"	6'7"	10'3"	8'4"	12'6"	10'2"	14'6"	11'10"
Douglas Fir-Larch	#1	7'10"	6'4"	11'5"	6'7"	14'5"	11'9"	17'8"	14'5"	20'5"	16'8"
	#2	7'5"	6'0"	10'10"	8'10"	13'8"	11'2"	16'9"	13'8"	19'5"	15'10"
	#3	5'8"	4'7"	8'3"	6'9"	10'6"	8'7"	12'9"	10'5"	14'10"	12'1"
Hem-Fir	#1	7'8"	6'3"	11'3"	9'2"	14'3"	11'8"	17'5"	14'3"	20'2"	16'6"
	#2	7'2"	5'10"	10'6"	8'7"	13'4"	10'10"	16'3"	13'3"	18'10"	15'5"
	#3	5'6"	4'6"	8'1"	6'7"	10'3"	8'4"	12'6"	10'2"	14'6"	11'10"

2015 IRC®, International Code Council®